LARGE-SCALE REGIONAL WATER RESOURCES PLANNING

Water Science and Technology Library

VOLUME 7

The titles published in this series are listed at the end of this volume.

LARGE-SCALE REGIONAL WATER RESOURCES PLANNING

The North Atlantic Regional Study

by

DAVID C. MAJOR

New York City Department of Environmental Protection,
Bureau of Water Supply and Wastewater Collection, New York, U.S.A.

and

HARRY E. SCHWARZ

Clark University, Worcester, U.S.A.

KLUWER ACADEMIC PUBLISHERS

DORDRECHT / BOSTON / LONDON

Library of Congress Cataloging in Publication Data

Major, David C.
Large-scale regional water resources planning : the North Atlantic regional study / by David C. Major and Harry E. Schwarz.
p. cm. -- (Water science and technology library ; 7)
Includes bibliographical references.

1. Water resources development--Atlantic Coast Region (U.S.)
I. Schwarz, Harry E. II. Title. III. Series.
TC423.1.M35 1990
333.91'15'0974--dc20 90-4213

ISBN 978-94-010-7401-8 ISBN 978-94-009-2011-8 (eBook)
DOI 10.1007/978-94-009-2011-8

Published by Kluwer Academic Publishers,
P.O. Box 17, 3300 AA Dordrecht, The Netherlands.

Kluwer Academic Publishers incorporates
the publishing programmes of
D. Reidel, Martinus Nijhoff, Dr W. Junk and MTP Press.

Sold and distributed in the U.S.A. and Canada
by Kluwer Academic Publishers,
101 Philip Drive, Norwell, MA 02061, U.S.A.

In all other countries, sold and distributed
by Kluwer Academic Publishers Group,
P.O. Box 322, 3300 AH Dordrecht, The Netherlands.

Printed on acid-free paper

"As planners it is your job to illuminate the choices."

Dr. Abel Wolman

TABLE OF CONTENTS

Preface

This is a book on the application of large-scale regional water resources planning methods in the North Atlantic Regional Water Resources Study (NAR), a study completed in 1972 for the United States Water Resources Council. This framework study, designed to serve as a guide for subsequent detailed basin and project planning in the region, took place following the development of new methods of water planning in the 1960's and is notable for the integration of these into the planning process. The study is important for its contributions to multiobjective (including the environmental objective) planning, for the use of mathematical models, and for the development of institutional and organizational measures to shape the planning process. The aim of this book is to make accessible to professionals and advanced students the lessons of what we believe to be a landmark study in the development of water planning methods.

The book is based on the 25 volumes of the study report, on other study drafts and publications, on articles written about the study, and on our own reflections on the study experience. The work discussed in this book represents the combined efforts of many professionals. The NAR is their accomplishment and we are grateful to them.

We are grateful in particular to the successive Division Engineers of the North Atlantic Division, Corps of Engineers, who provided both the leadership and the freedom for innovative work that made the NAR possible. They were, in chronological succession: Brigadier General David S. Parker; Brigadier General Harry G. Woodbury; Brigadier General Francis P. Koisch; Major General Charles M. Duke; and Major General Richard H. Groves.

We owe a great debt to the distinguished members of the study's Board of Consultants, who encouraged us to use new techniques and who acted as our collective conscience. The members of the Board, with their positions at the time of the study, were: Mr. David V. Auld, Consulting Engineer and former Director of Sanitary Engineering for the District of Columbia; Dr. Arthur Maass, Professor of Government, Harvard University; Mr. Eugene W. Weber, Consultant and former Deputy Director, Civil Works, Corps of Engineers; Dr. Gilbert F. White, Professor of Geography, University of Colorado; Dr. Nathaniel Wollman, Dean, College of Arts and Sciences, University of New Mexico; and Dr. Abel Wolman, Professor Emeritus of Sanitary Engineering, Johns Hopkins University. The injunction following the title page was given to us at the first meeting of the NAR Coordinating Committee by the late Dr. Wolman, our role model, advisor and friend.

We are grateful also to the members of the Coordinating Committee and their alternates. The 84 professionals who served in these capacities during the course of the study worked hard and well in an innovative institutional structure.

We wish to thank also the members of our own staff at the North Atlantic Division, U.S. Army Corps of Engineers, New York, and the staffs of the cooperating States and agencies who worked with them. More than 200 Corps, State, and agency personnel contributed full or part time to the study.

We acknowledge also the contributions to the study of an exceptional group of contractors: Research Planning and Design Associates, Amherst, Mass.; Cornell University, Ithaca, N.Y.; Regional Plan Association, New York, N.Y.; The Travelers Research Corporation, Hartford, Conn.; Peter S. Eagleson, Consulting Engineer, Cambridge, Mass.; Meta Systems Inc., Cambridge, Mass; and Ralph M. Field, Planning Consultant, New York, N.Y.

For reading the manuscript in whole or in part, we are indebted to Eugene Fleming, Arthur Maass, Peter Rogers, Ira Silver, Abel Wolman and Ervin Zube.

Major wishes to record his gratitude to the family of Phyllis Ann and Stephen A. Hart for their hospitality during the writing of the book, and to the President and Fellows of Clare Hall, The University of Cambridge, who extended to him a Visiting Fellowship during which the book was completed. Schwarz thanks his colleagues at Clark University for their understanding during the many hours when he was hidden away from university duties, and his wife Elizabeth for her patience on evenings and weekends.

The authors also wish to acknowledge their happy professional collaboration of 25 years, beginning with the participation of both in the Harvard Water Program and continuing with their work as Branch Chief and Chief Planner (Schwarz) and Chief Economist (Major) for the North Atlantic Regional Water Resources Study. This book is affectionately dedicated to our spouses, Patricia Hart and Elizabeth Schwarz.

Harry E. Schwarz
Worcester, Mass.

David C. Major
New York, N.Y.

Introduction

The North Atlantic Regional Water Resources Study (NAR) was completed in 1972 by the United States Army Corps of Engineers, North Atlantic Division, for the North Atlantic Regional Study Coordinating Committee, a Federal-State body chartered by the United States Water Resources Council. The study is important for its contributions to multiobjective planning, including the environmental objective; for the use of mathematical modeling methods; and for the use of institutional and organizational measures for planning. As these three elements define in substantial measure today's best-practice water resources planning, their use in the NAR accounts for both the study's contemporary relevance and its historical importance. The purpose of this book is to present the methods of the study and perspectives on them in order to illustrate the lessons of the study for professional planners and students of water resources planning and regional planning more generally.

The goals of the North Atlantic study were those of the "framework" studies of water and related land resources of the United States Water Resources Council. These regional planning studies, of which some 20 were envisioned and more than half ultimately completed, were designed to examine water resources needs and supplies by regions, thereby serving as guides to detailed basin and project planning efforts and assuring a measure of consistency among them. The studies were to examine needs and supplies by river reaches or other subregions, rather than on a project level. The NAR planners, using this charter as a basis, went beyond it in the use of multiobjectives, mathematical models, organizational measures for planning, and in other ways.

THE BOOK AND THE STUDY

This book is intended to describe primarily those NAR planning methods that were new: the combination of multiobjective methods, mathematical models, and planning processes that made the study an innovative one. Some of the detailed techniques used to produce inputs to the overall planning process were also new (in part at least because all of the work was done within a new context) and many of these are described here. On the other hand, some of the techniques used to produce required inputs were standard methods of economics, engineering, and water planning. Most of these are not discussed in this book; they are explained in the detailed subject appendices to the NAR report. With the exception of descriptions of such standard material, the book is designed to give a comprehensive view of the NAR study.

The book is organized into three parts. Part I, Chapters 1 and 2, provides the background to the study: the context of water planning at the time of the study, the region as seen by the planners, and an overview of the NAR methods. Part II, Chapters 3-8, describes the applications of the methods: multiobjectives, the demand, supply, and storage-yield models, and institutional arrangements. Each of the chapters in Parts I and II includes a section that provides perspectives on the methods. Part III, Chapters 9 and 10, describes the outputs of the study and presents perspectives on the study as a whole.

USING THE BOOK IN TEACHING

The book makes available to teachers and students material that we have found to be of value in our teaching at, successively, the Massachusetts Institute of Technology and Clark University, as well as in our professional work as water resource planners. We expect that the book will be of use in advanced courses in water resources planning, in resource analysis and in planning more generally. In such courses, we have generally used the NAR material in the chapters of the book as examples of particular planning applications: multiobjectives, including the visual and cultural analysis of Chapter 4; the demand (input-output), supply (linear programming) and storage-yield models in Chapters 5, 6 and 7, and the material on institutional measures in Chapter 8. These examples fit in well with the material in such often-used works in the field as Maass et al. (1962), Marglin (1967), UNIDO (1972), Major (1977), Cohon (1978), and Goodman (1984).

The NAR study is also appropriate as the basis of an advanced planning seminar such as Geography 222, River Basin Planning, a seminar given at Clark every other year. In this seminar, a particular river basin plan or regional plan is examined in the light of planning theory, and student papers are written on particular aspects of the plan under consideration. Each student is asked to take an aspect of the plan, for example the NAR's visual quality evaluation procedures, and to evaluate its relative success and to suggest alternatives or improvements. A more ambitious undertaking would be to compare and contrast the whole methodology and procedures of the NAR with those of a more traditional basin plan (for example, the Delaware River Basin Report, U.S. Army Corps of Engineers, 1962) or with other advanced planning efforts, such as that for the Rio Colorado, Argentina, described in Major and Lenton (1979).

NAR REPORTS

The principal reports of the NAR study are listed in Table 1. A bibliographical reference for the study reports is: United States North Atlantic Regional Water Resources Study Coordinating Committee, North Atlantic Regional Water Resources Study, Report, Annexes, and Appendices, North Atlantic Division, U.S. Army Corps of Engineers, New York, 1972. A complete list of all of the draft reports, final reports, and consultants' reports developed for the project is given in Appendix A of the study, pp. A-34 to A-45. Table 2 of this introduction shows the reports on which individual chapters of the book are principally based. Articles written on the NAR study include: Schwarz and Major (1971); Schwarz (1972); Schaake and Major (1972); deLucia and Rogers (1972); Major (1972, 1973); and Schwarz, Major and Frost (1975). We have drawn on these as well as on the project reports in the preparation of this book.

TABLE 1

Principal Reports of the NAR Study

Main Report

Annex 1: Area Summaries
Annex 2: State Summaries

Appendices

A. History of Study
B. Economic Base
C. Climate, Meteorology, Hydrology
D. Geology and Groundwater
E. Flood Damage Reduction
F. Upstream Flood Prevention
G. Land Use and Management
H. Minerals
I. Irrigation
J. Land Drainage
K. Navigation
L. Water Quality and Pollution
M. Outdoor Recreation
N. Visual and Cultural
O. Fish and Wildlife
P. Power
Q. Erosion and Sedimentation
R. Water Supply
S. Legal and Institutional Environment
T. Plan Formulation
U. Coastal and Estuarine Areas
V. Health Aspects

Selected Other Reports

Plan of Study, Vols. I, II (cited as PS I, II)
Special Publication: Manuals for Computer Programs of NAR Study: Storage-Yield, Demand and Supply Models (cited as SP)

TABLE 2

NAR Reports on which Chapters of the Book are Principally Based.

Chapter	Reports
1	Report, Appendix A
2	Report, Appendices A and T
3	Report, Appendix T
4	Appendix N
5	Appendix T, Special Publication
6	Appendix T, Special Publication
7	Special Publication
8	Appendix A and T
9	Report, Annex 1 and 2
10	Report

AVAILABILITY OF THE NAR REPORTS

Of the many copies of the NAR reports published under the auspices of the Coordinating Committee, most have found their way into inaccessible agency storage or are otherwise unavailable. For the convenience of professionals and students who wish to consult the original reports to explore further the ideas and methods described in this book, we have deposited a complete set of the reports listed in Table 1 with the Robert S. Goddard Library, Clark University, Worcester MA 01610 USA. Readers wishing to consult these should contact the Goddard Library. In addition, the NAR reports are available on an interlibrary loan basis from the Headquarters Library, U.S. Army Corps of Engineers, 20 Massachusetts Avenue N.W., Washington DC 20314-1000 USA.

REFERENCE STYLE

Two reference styles are used in the book. The first is used for all references except those to the NAR reports, and the second is used for references to the reports.

The first style is standard journal style. The name of the author or authors of a document, the date of publication, and chapter or page references as appropriate are given. Thus: Maass et al. (1962, p. 17). A full bibliographic citation for each document referenced in this style will be found in the reference list at the end of each chapter in which reference is made to the document.

The second reference style is an abbreviated style, used because of the frequency with which reference is made to the NAR reports. This reference style includes a short title and page reference within parentheses. Thus: (Report, 3) and (Annex 1, 15-16), refer respectively to the main report of the study, page 3, and to the first annex to that report, pp. 15-16. References to appendices to the main report, each of which was given a letter as well as a title by the project staff, are further compressed. Thus: (T-155) refers to appendix T, Plan Formulation, p. 155, and (SP-224) refers to the Special Publication, p. 224. All of the NAR reports and documents to which reference is made are included in the list of reports given in this chapter. The NAR reports referenced in a given chapter are not included in the reference list at the end of that chapter.

PARTICIPANTS IN THE PROJECT

A large number of persons participated in the NAR, on the Coordinating Committee, the Board of Consultants, in the U.S. Army Corps of Engineers, North Atlantic Division and in the cooperating State and Federal agencies, and in the private firms associated with the study. The participants are listed on Report, 225-226 (the Coordinating Committee, the Board of Consultants, and the Executive Secretaries) and A-25 to A-34 (agency staff members).

UNITS

The NAR study employed a mix of traditional (English) and metric units. Table 3 shows the principal units used, their equivalents in the system not used, and common multiples or submultiples.

TABLE 3: UNITS OF MEASUREMENT EMPLOYED IN THE STUDY

Type of Measurement	Unit	Equivalent	Multiples or Submultiples
Length	mile	1 mile=1.609 km	1 mile=5280 ft
Area	acre	1 acre=0.405 ha	1 acre=43,560 ft^2
Volume	acre-foot	1 af=1234 m^3	
Flow	cfs	1 cfs=0.0283 m^3/sec	
Power	MW	1 MW=56880 Btu/min	1 MW=1,000 kw
Energy	GWh	1 GWh=3.412×10^9 Btu	1 GWh=10^6 kwh
Temperature	^oF	$^oF = 9/5^oC+32$	

REFERENCES

Cohon, Jared, Multiobjective Programming and Planning, Academic Press, New York, 1978.

deLucia, Russell J., and Peter Rogers, "North Atlantic Regional Supply Model," Water Resources Research 8:3, June, 1972, 760-765.

Goodman, Alvin S., Principles of Water Resources Planning, Prentice-Hall, Englewood Cliffs, N.J., 1984.

Maass, Arthur, et al., Design of Water-Resource Systems, Harvard University Press, Cambridge, Mass., 1962.

Major, David C., "Impact of Systems Techniques on the Planning Process," Water Resources Research 8:3, June, 1972, 766-768.

Major, David C., "Water Planning Models in the North Atlantic Study," Journal of the Urban Planning and Development Division 99:UP2, September, 1973, 131-136.

Major, David C., Multiobjective Water Resources Planning, American Geophysical Union, Washington, D.C., 1977.

Major, David C. and Roberto L. Lenton, Applied Water Resource Systems Planning, Prentice-Hall, Englewood Cliffs, N.J., 1979.

Marglin, Stephen A., Public Investment Criteria, MIT Press, Cambridge, Mass., 1967.

Schaake, John C. Jr. and David C. Major, "Model for Estimating Regional Water Needs," Water Resources Research 8:3, June, 1972, 755-759.

Schwarz, Harry E. and David C. Major, "An Experience in Planning: The Systems Approach" Water Spectrum 3:3 Fall, 1971, 29-34.

Schwarz, Harry E., "The NAR Study: A Case Study in Systems Analysis," Water Resources Research 8:2, June 1972, 751-754.

Schwarz, Harry E., David C. Major and John E. Frost Jr., "The North Atlantic Regional Study," in J. Ernest Flack, ed., Proceedings of the Conference on Interdisciplinary Analysis of Water Resources Systems, American Society of Civil Engineers, New York, 1975, 245-271.

UNIDO (United Nations Industrial Development Organization), Guidelines for Project Evaluation, United Nations, N.Y., 1972.

U.S. Army Corps of Engineers, Delaware River Basin, New York, New Jersey, Pennsylvania and Delaware, H. Doc. 87-522, 11 vols., August 16, 1962.

PART I: BACKGROUND, DESCRIPTION, METHODOLOGY AND OVERVIEW

Chapter 1: The Study and the Region

This chapter describes the study and the region: the context of water resources planning within which the study took place and the developments in water resources planning that followed; the authority and history of the NAR study; the physical, demographic and economic characteristics of the region as seen by the planners; and perspectives on data.

1.1 THE CONTEXT OF WATER RESOURCES PLANNING AT THE TIME OF THE NAR STUDY AND DEVELOPMENTS THAT FOLLOWED

The NAR study took place during a major shift in theoretical approach toward water resources planning, namely the introduction of multiobjective planning; and during a time of increasing technical capability in terms of digital computer models for water planning. The study was shaped by these influences, and also had an impact on them, particularly on the first in the form of U.S. Water Resources Council (1973) criteria for planning.

The development of water planning criteria in explicit form prior to 1962 was in essence the story of the application of a popular model of "welfare" economics (Bergson, 1938; Graaff, 1963) to water resources planning. This model has often been interpreted to mean that, under certain restrictive assumptions, public expenditure might suitably be directed toward the maximization of the net discounted national income benefits of projects (Major, 1977, ch. 1). Perhaps because of its relative newness and an attractive analytical rigor, the divergence of the model's assumptions (and thus its policy guidance) from reality was not a preoccupation of planners during the early post World War II years. The application of a one dimensional policy guide to multi-dimensional social decision problems such as those presented in most water resources planning investigations was bound to give rise to a tension between the narrowly focussed tool and the broadly focussed or multiobjective social decision. This tension was explored in most detail, and the results of the exploration most effectively presented, in the work of the Harvard Water Program, notably in Maass et al., (1962). The publication of Design of Water-Resource Systems was a turning point in water planning away from the efficiency-oriented approach epitomized in the "Green Book" (U.S. Inter-Agency River Basin Committee, 1950, rev. 1958) toward multiobjective planning. The NAR study was the first planning study to use the multiobjective approach for water resources in a thoroughgoing manner. The publication of Design of Water-Resource Systems also presented effectively the use of both simulation and mathematical programming models in water planning.

This presentation was part of the general response of water planning professionals to the growing availability of computational power, and it directly affected the NAR study, which was innovative with respect to models as well as with respect to multiobjectives.

The time coextensive with the study and immediately following it saw the development of explicit multiobjective regulations to govern Federal water resources planning in the United States, the Principles and Standards of 1973 (U.S. Water Resources Council, 1973). (An insightful assessment of these and later criteria is given in Stakhiv (1986). The cost-sharing and local participation provisions of the Water Resources Development Act of 1986 (P.L. 99-662) together with existing environmental legislation insure that Federal water resources planning remains multiobjective.) The period during and since the study has also seen a flowering of modeling studies for water resources planning. The NAR study is thus a precursor with respect to applications of the principal lines of approach to modern water resources planning, and in many ways its methods remain models for current use. This holds for the institutional methods of the study, described in Chapter 8, as well as for the multiobjective and modeling methods, described in Chapters 3, 4, 5, 6, and 7.

1.2 AUTHORITY AND HISTORY

The NAR study was authorized in P.L. 89-298 of October 27, 1965, section 208. The Corps of Engineers was designated as the interdepartmental lead agency for the NAR, and the Chief of Engineers assigned the study to the Division Engineer, North Atlantic Division (A-1). The NAR study was one of some 20 framework studies for the United States that were scheduled to be completed for the U.S. Water Resources Council. More than half of these studies, including the NAR, were completed before the program was superseded by a national assessment program.

The general purposes of the NAR, as interpreted by the study team, were to estimate demands for and supplies of water resource system outputs on a broad scale (river reaches, for example); to point out areas for priority basin and project studies; and to highlight areas of needed research. The framework studies were not, in general, to deal with individual projects; project studies were to be included in basin and project planning efforts undertaken within the general guidelines provided by the framework studies. The geographic boundaries of the NAR, as described in Section 208 of P.L. 89-298, included all or parts of thirteen states and the District of Columbia (see Figure 1-1). The region includes about 5% of the nation's land area and much larger proportions of its population and economic activity. The final NAR reports were transmitted to the Water Resources Council on 30 June 1972. The total planning time was six and one half years; planning activities were undertaken, informally at least, from the beginning of calendar 1966: in January of that year an ad hoc subcommittee for the

FIGURE 1-1 GEOGRAPHIC BOUNDARIES OF THE NAR

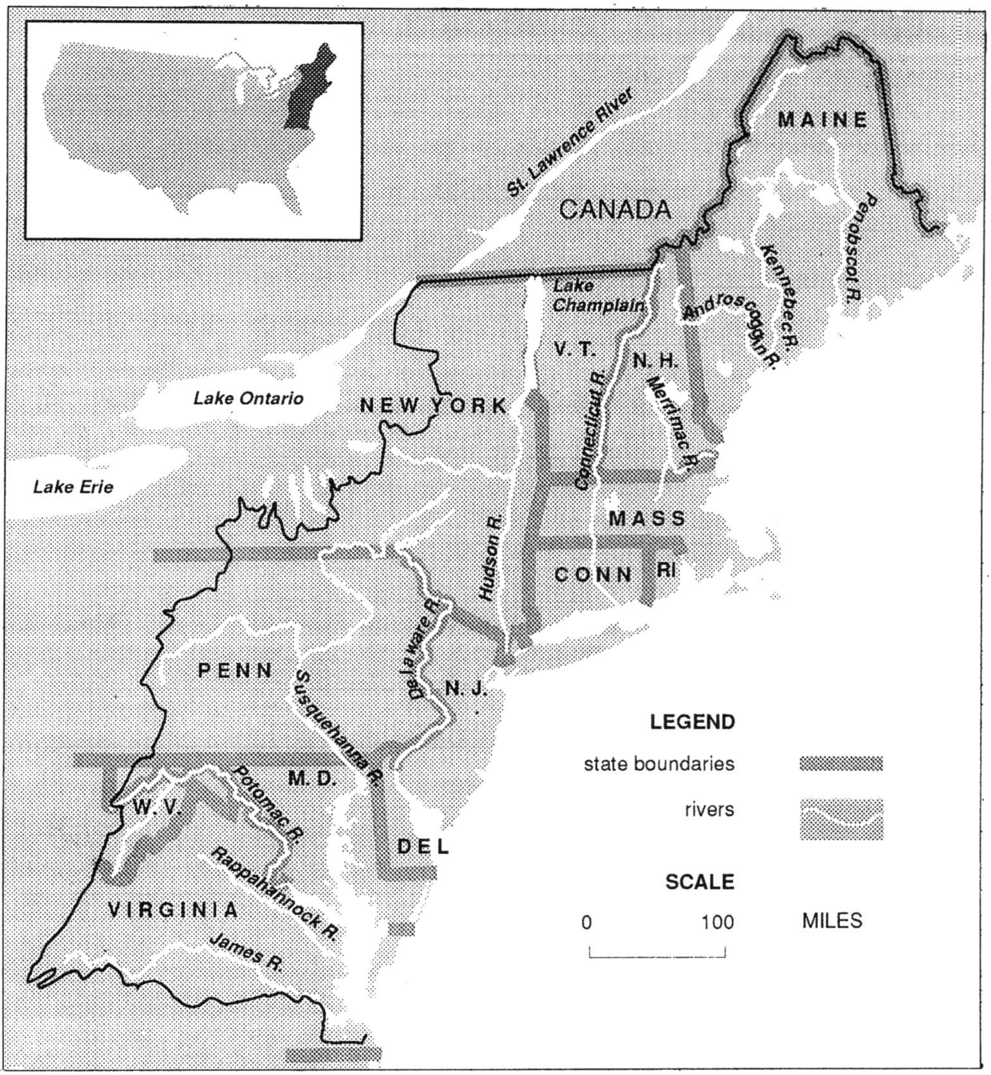

plan of study was formed (A-10). The total funds (the sum of current dollars for the study years) appropriated for the study were $4,683,000 (A-20).

1.3 PHYSICAL, DEMOGRAPHIC AND ECONOMIC CHARACTERISTICS OF THE REGION

This section describes the physical, demographic and economic characteristics of the NAR region as seen at the time of planning; these characteristics provide the general framework with reference to which the planners made forecasts of needs, sources and devices for water resources development and management described later. The material in this section is from Report 18-48; it is presented here in paragraphs on physical characteristics; precipitation and water supply; population; economy; and institutions.

1.3.1 PHYSICAL CHARACTERISTICS. The North Atlantic Region stretches along the Atlantic Coast from the southern part of Virginia to the northern tip of Maine. It extends inland to encompass all of the land drainage which flows into the Atlantic Ocean through this coastal zone. All or portions of 13 states and the District of Columbia are in the region, which measures approximately 1,000 miles along a general north-south axis and averages 200 miles in width. The region contains 172,586 square miles of land and water, about 5 % of the total national area.

The topography of the NAR varies from mountainous terrain with elevations over 4,000 feet to flat and undulating coastal plains. Proceeding westward from the Atlantic coast, the topography ranges from coastal plains through undulating hills to rolling hills which generally have elevations of from 200 to 800 feet, to steep hills of from 800 to 2,000 feet, and finally to the Appalachian mountains in the westernmost part of the region. The predominant land form is rolling hills: these make up more than half of the NAR's land area.

Forest cover was, at the time of planning, by far the largest land use in the NAR, with about 66 percent of the total acreage, followed by cropland with 15 percent, urban and "other" (roads and open tracts) land uses with 13 percent, and pastureland with 6 percent.

1.3.2 PRECIPITATION AND WATER SUPPLY. The average annual precipitation in the NAR is about 41 inches. The range is from less than 30 inches in northern New York to over 70 inches in some mountainous areas. The distribution of precipitation is relatively uniform throughout the year in all parts of the NAR except the coast. The coast receives greater autumn precipitation, less in winter and spring, and about the same in summer as compared to the rest of the region.

Within the region the major river systems include the Connecticut, Hudson, Delaware, Susquehanna, Potomac and James. All of the surface and groundwater resources of the region combine to provide an abundant supply of water, although one that is somewhat unevenly distributed geographically and degraded in quality in some locations, particularly in metropolitan areas.

Streamflow from the region, including the flow originating in Canada, averages about 260,000 cfs, or about 9 percent of the annual natural runoff of the United States. Since the region represents about 5 percent of the nation's area, the average flow on a per square mile basis is nearly twice the national average. The average streamflow within the NAR is about 1.5 cfs per square mile, varying from 2.5 cfs per square mile in a few locations in New England and New York to less than 1 cfs per square mile throughout much of the Potomac, York and Rappahannock river basins. This supply is not naturally distributed throughout the region for convenient availability at locations of greatest need. Moreover, the use of streamflow is limited by large annual and in some areas seasonal variations in precipitation and runoff. (The effects of these variations were demonstrated during the droughts of the 1960's and the 1980's.) However, extended periods of low flow, such as occur during droughts, are relatively infrequent in the NAR and seasonal variations play a more immediate role in many water problems of the region.

There were at the time of planning almost 16 million acre-feet (maf) of existing usable storage in reservoirs with volumes larger than 5,000 acre-feet. Northern New England contained the greatest amount of this large storage, with about 4.8 maf. Much of the storage in the northern part of the region is used for power, while storage in the central and southern portions is used relatively more for flood control and municipal supply. Additional storage is available in numerous smaller upstream reservoirs. The release from reservoirs in many basins of the northern part of the region has a significant effect on dry weather flow, reducing stream flow variations considerably.

Surface water accounts for about 4 percent of the total land and water area of the region. This figure includes natural and artificial lakes, ponds, streams, estuaries and canals one-eighth of a mile wide or wider and deeply indented embayments, sounds and other sheltered coastal waters protected by headlands or islands. Half of this water surface, nearly 3,300 square miles, is in New England. The overall quality of surface water in the region varies widely with location, and is generally poorer at times of low flow.

Groundwater is in generally abundant supply with available rates of flow of up to several tens of millions of gallons per day in some well fields in the coastal plain, in sandstone and carbonate rocks, and in glacial sands and gravel beds in the more northerly areas. Smaller quantities of ground water are available throughout the region in places underlain by crystalline rock and shales, with yields limited to about 1 mgd per well field.

Groundwater development in 1965 amounted to about 2,560 mgd, with about 45 percent used for municipal supply, 15 percent for rural domestic supply, 34 percent for self-supplied industrial use, 3.5 percent for irrigation and the remainder for electric power and livestock. New York State was the largest user with about 600 mgd, more than half on Long Island. (A more recent estimate of aggregate groundwater withdrawals on Long Island is 486.2 mgd: New York State Department of Environmental Conservation, 1982, p. 2.) New Jersey was the next highest user, with withdrawals of about 530 mgd.

Groundwater quality on an overall regional basis was good. Hardness in the region is encountered in carbonate rock and iron or manganese are found in undesirable quantities in glacial deposits. Wells in coastal plain sediments near the sea may yield water that is high in chloride content, limiting its use.

1.3.3 POPULATION. The region was and is characterized by varying population densities ranging from the lightly populated wildlands and wilderness areas of northern Maine to the heavily concentrated center-city and urban "megalopolis" stretching from Boston to Washington, D.C. and beyond (see Figure 1-2). This "megalopolis" of urban and suburban development occupied at the time of planning between 30 and 40 percent of the region's area and contained between 81 and 86 percent of its population, depending on the definition of the "megalopolis." The planners expected that the population of the NAR would nearly double by the year 2020 to 86,199,800 persons. The rate of growth was forecast to be about four-fifths of that projected for the country as a whole, and as a result the NAR's share of the projected national population in 2020 would decline to about 22 percent. This total population would become even more predominately urban than it was, and urban areas would grow in physical size. However, it was not anticipated that central city population densities would substantially increase, but rather that population growth within the urban belt would occur mostly in the suburbs, with some of these developing into new urban concentrations.

1.3.4 ECONOMY The NAR plays an important role in the nation's economy. The region in 1960 contained 26 percent of the nation's population and received about 30 percent of the country's total personal income. Among the most important economic activities in the region are manufacturing (including manufacturing in such water-using industries as textiles, chemicals and paper products), Federal government activities, and the finance and service industries. Extractive activities (mining, agriculture, forestry, and fisheries) play a relatively small role. The importance of population-related activities such as trade, transportation, and construction is, on a relative basis, about the same in the NAR as in the nation as a whole.

Since manufacturing, government, finance and services are relatively labor-intensive, they tend to be localized in and around the densely populated urban belt of the region. Examples include electronics in

FIGURE 1-2 POPULATION DENSITIES IN THE NAR (Source: Report, 40)

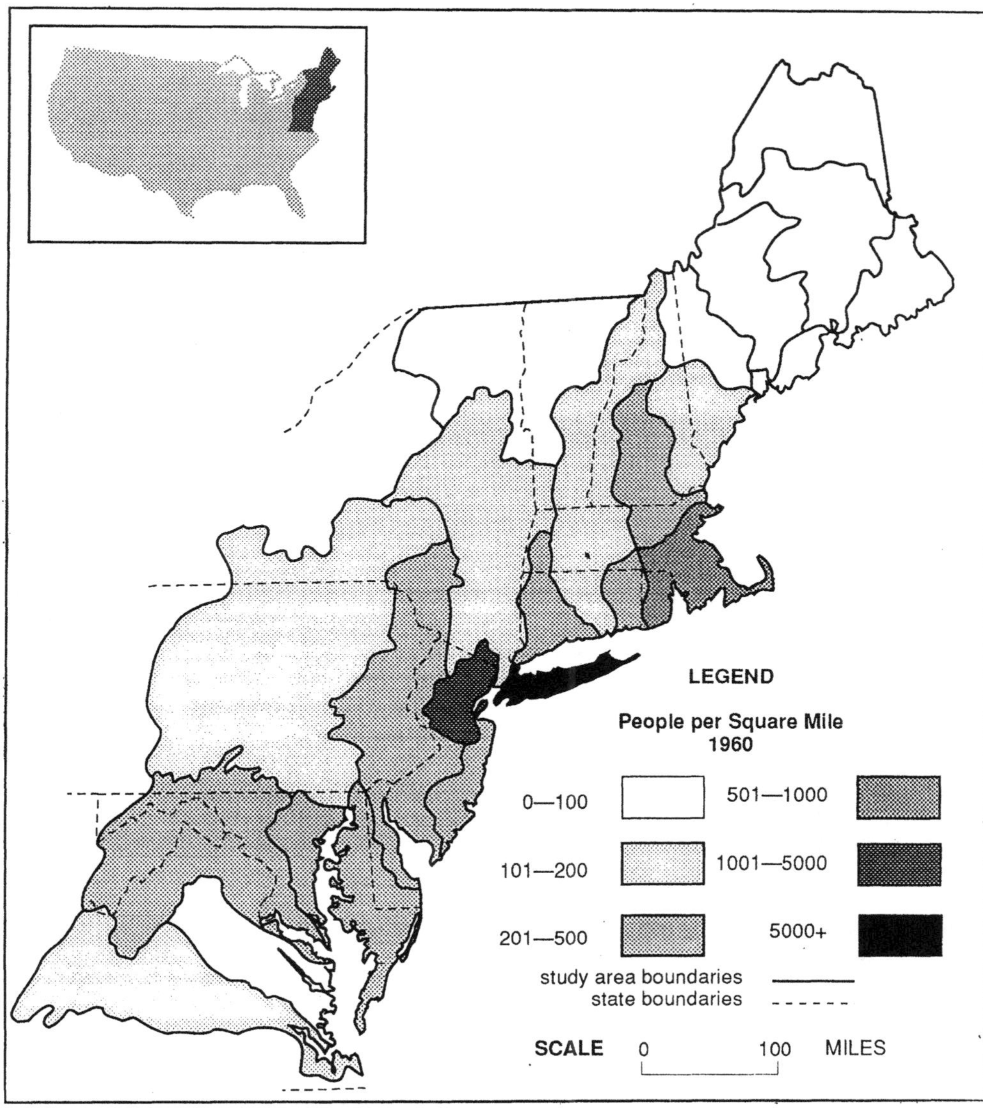

the Boston area, the finance, publishing, and garment industries in New York City, petrochemicals in northern New Jersey, and government in Washington, D.C.

Land and resource intensive agriculture, fishing, forestry, and mining are important in the less populated exurban areas, where land is not at a premium or where scarce resources are located. Examples include coal mining in the Susquehanna and Potomac River basins and commercial fin and shell fisheries in the Chesapeake Bay estuarine system.

At the time of planning the NAR was growing at a slower rate than the nation as a whole, as measured by population, employment, and income. This growth rate differential was projected to continue to 2020.

The urban belt of the NAR was expected to grow both in area and in share of the region's total population over the study period. This change is concurrent with migration within the urban belt from central cities to suburbs. A swiftly growing strain on land use management practices in the suburbs was expected as a result of such population changes, along with increasing pressure on water supply facilities, sewage disposal and waste treatment plants and electric power generating capacity. All of these changes directly and indirectly affect major aspects of water and related land resources planning.

The total demand for water in manufacturing was forecast to increase although the relative importance of this in the NAR was expected to decline moderately. Manufacturing activity was likely to shift from central cities to suburbs, both to follow the labor force and to expand on less expensive land. Agriculture, forestry, fishing and mining were projected to decrease in relative (and, in some cases absolute) importance as land values are pushed upward by the spread of urbanization and industrialization. However, the planners saw an element of uncertainty here. For example, changes in energy technology could either introduce new vigor into the coal extraction industry or render it obsolete. Then, too, there are untapped mineral reserves on the continental shelf whose exploitation was held to depend on the resolution of opposing pressures of economics and environmental protection.

The economic base of the region as a whole, and of the urban areas in particular, was expected to evolve in the direction of the "office" activities of government, finance, and services. As in the case of manufacturing, employment in these "office" occupations was projected to shift outward to the suburbs, a shift that has since been reflected in the development of substantial suburban office centers in the region.

Land and water characteristics of the region have not served in general to limit economic development, and were not expected to do so in the future. The NAR as a whole is well endowed with water, and with certain exceptions, droughts have been comparatively mild and localized. The uneven geographic distribution of water poses region-wide problems of management and allocation. Although the NAR is the most urbanized region in the nation, its urban belt occupies less than 40 percent of its land area, and it seemed evident to the planners that the future economy of the region would be shaped more by the decisions of man than by the restrictions of nature on land and water supplies. The planners did not consider long-term climatic fluctuations and the impact of human activities on these (for example, the "greenhouse effect"), phenomena that have more recently become of substantial concern and that may have significant impacts on regional economic vitality. For an example of the appreciation of the possible impacts of long term climatic fluctuations on water planning, see New York City (1987, p. 17).

1.3.5 INSTITUTIONS. In part because the NAR is well endowed with water, a comprehensive body of case law has not been generated to resolve water rights controversies. The common law rule, with its variety of interpretations, still stands; a riparian owner has the right to reasonable use of water which flows by his property. Perhaps more important for water management in the NAR are the volume of statutes governing water use and the number of agencies that administer water under these statutory guidelines. These statutes, the source of authority for the agencies with which the local planner deals, formed part of the basis for interagency and intergovernmental cooperation in the NAR study. Institutional arrangements for planning the development of water and related resources in the NAR encompass each of the many levels of government, and each of the agencies which executes and administers the laws governing those resources. These levels include Federal, state, local, regional, river basin and interstate agencies.

1.4 PERSPECTIVES ON DATA

Substantial amounts of information on the physical, demographic, economic and institutional characteristics of the region were available at the start of the NAR study. This information was taken from a wide range of government and private documents including many previous water resources studies in the region. Framework studies can be based on comparable information almost everywhere in the developed world, and much of this information is now available in electronic form. However, there will normally be a need for some additional data collection such as, in the NAR study, the landscape evaluation data described in Chapter 4 or the groundwater cost estimates used in modeling (Cederstrom, 1973). In developing areas, more substantial data collection efforts will often be required.

To use planning resources efficiently, data collection efforts need to be carefully designed. The proper allocation of resources to data collection and organization is a critical part of planning for a framework study. Data should be sufficient for the decisions to be made on the basis of the study; excessive or unorganized data collection dissipates time and resources. (For a discussion of mathematical modeling techniques and their relationship to data, see Major and Lenton, 1979, ch. 3.) For example, in order to estimate the cost and yield relationships required in framework supply modeling, a general survey of storage potential in a region will often suffice, obviating the need to allocate resources to the detailed assessment of specific reservoir sites. The developing computer-based geographic information systems may well become helpful tools in framework planning (Brown and Gersmehl, 1987).

REFERENCES

Bergson, A., "A Reformulation of Certain Aspects of Welfare Economics," Quarterly Journal of Economics 52:2, 1938, 310-334.

Brown, D.A. and P.J. Gersmehl, eds., File Structure Design and Data Specifications for Water Resources Geographic Information Systems, Water Resources Research Center, University of Minnesota, Minneapolis, Minn., Report No. 10, 1987.

Cederstrom, D.J., "Cost Analysis of Ground-Water Supplies in the North Atlantic Region, 1970," U.S. Geological Survey, Water-Supply Paper 2034, 1973.

Graaff, J. de V., Theoretical Welfare Economics, Cambridge University Press, New York, 1963.

Major, David C., Multiobjective Water Resource Planning, American Geophysical Union, Washington, D.C., 1977.

Major, David C. and Roberto L. Lenton, Applied Water Resource Systems Planning, Prentice-Hall, Englewood Cliffs, N.J., 1979

Maass, Arthur, et al., Design of Water-Resource Systems, Harvard University Press, Cambridge, Mass., 1962.

New York City, Mayor's Intergovernmental Task Force on New York City Water Supply Needs, Managing for the Present, Planning for the Future, Second Interim Report, New York, December, 1987.

New York State, Department of Environmental Conservation, Division of Water, Water Supply Unit, "Report of Long Island Groundwater Withdrawal during 1981," Albany, N.Y., December, 1981.

Stakhiv, Eugene Z., "Achieving Social and Environmental Objectives in Water Resources Planning: Theory and Practice," in Warren Viessman Jr., and Kyle E. Schilling, eds., Social and Environmental Objectives in Water Resources Planning and Management: Proceedings of an Engineering Foundation Conference, Santa Barbara, California, May 11-16, 1986, American Society of Civil Engineers, New York, 1986, 107-125.

U.S. Inter-Agency River Basin Committee, Subcommittee on Benefits and Costs, Report to the Federal Inter-Agency River Basin Committee: Proposed Practices for Economic Analysis of River Basin Projects, rev. ed., Washington, D.C., 1958 ("The Green Book," originally published 1950).

U.S. Water Resources Council, "Water and Related Land Resources: Establishment of Principles and Standards for Planning," Federal Register 38: 174, 1973, 24778-24869.

Chapter 2: NAR Planning: An Overview

This chapter provides an overview of the planning methods used in the NAR study. First, the principal methods used in planning the NAR are outlined: multiobjectives; mathematical models; and institutional arrangements. Then, three other elements of the planning process are described: the analysis of needs, sources and devices; the planning regions; and the forecast years. A final section contains perspectives on the methods. The chapter provides a framework for detailed treatments of the principal planning methods in the chapters to follow.

2.1 MULTIOBJECTIVES

The use of multiobjectives (Chapters 3 and 4) is one of the distinguishing features of the NAR study, along with the use of mathematical models and the institutional methods for planning that were utilized. The NAR was one of the first explicitly multiobjective Federal studies; its use of these techniques followed closely upon the publication of the landmark work on them, Maass et al. (1962). Multiobjective analysis is a generalization of traditional cost-benefit analysis. Traditional analysis focuses on the national income (or "economic") objective of water planning. Multiobjective analysis, by contrast, emphasizes the design of water projects and programs in terms of all relevant objectives, environmental, regional, social, and others, including the national income objective. In practice, as in the NAR, multiobjective analysis is concerned with the choice of objectives, the development of alternative feasible plans responsive to objectives, and the final choice of a plan. Works that describe the theoretical structure of multiobjective planning include Maass et al. (1962); Marglin (1967); UNIDO (1972); Major (1977); and Major and Lenton (1979); see also Figure 3-1.

NAR planning took place in terms of three objectives: national income (the traditional objective of benefit-cost analysis); regional development; and environmental quality. These objectives, chosen for reasons described in Chapter 3, were formally adopted for NAR planning by the Coordinating Committee in September, 1967 (A-10).

The level of detail at which each multiobjective effect was examined in NAR planning varied depending on the importance of the effect for decision-making; the available staff, time and other resources; and the methods available. Relatively detailed work was done for economic cost accounting and for visual and cultural effects; on the other hand, economic benefits were discussed in terms of general magnitude, and regional effects were discussed rather broadly. The multiobjective approach was thus used in the NAR study as a conceptual framework to guide the organization and evaluation of information, rather than as a consistently detailed guide to the evaluation of individual projects. (For guidelines for the latter use see UNIDO, 1972.)

The approach to plan formulation for multiobjectives that was employed in the study included the development of three alternative plans, each with a somewhat exaggerated emphasis on a single objective. These would be examined by the Coordinating Committee, which would use them to decide on a final recommended mixed objective plan for the NAR. This approach to multiobjective plan formulation shaped both the models and the institutional methods used in the study.

2.2 MATHEMATICAL MODELS

Three models were important components of the NAR planning process: a demand (or requirements) model, based on input-output methods; a mathematical programming supply model; and a storage-yield model used to generate hydrologic inputs to the supply model. These models were utilized together to forecast requirements and to estimate costs of supply. The models were developed within the NAR planning context: a framework study; a commitment to multiobjective methods; large quantities of data from different agencies; and many standard methods of economics, engineering, and water planning. The three models are treated in detail in Chapters 5, 6, and 7 respectively. Two overviews of the use of models in water resources planning are provided in Friedman et al. (1984), and Rogers and Fiering (1986).

The demand model is a group of subroutines based on forecasting relationships designed to yield estimates of demands for water based on specific assumptions. The mathematical components of the model are an input-output table for the NAR region, a regression estimator for publicly supplied municipal and industrial water, and a group of arithmetic operations associated with these components. (References on input-output analysis, an economic simulation and forecasting method, are provided in Chapter 5.) The principal inputs to the model are projections of regional economic product, population, personal income, and water withdrawal coefficients, together with projections of the geographic distribution within the region of these variables. The model acts on the inputs to produce estimates by benchmark years of water flow demands by economic sector, by type of water quality, and by subbasin, basin, state or area. These flow demands are inputs to the supply model. Many different runs, responsive to alternative assumptions about forecasts and objectives, were made with this model; these are summarized and evaluated on T-271 to T-292.

The supply model is a mathematical programming (optimizing) model (see Chapter 6 for references to this technique). It is designed to permit evaluation of the sources and the costs of supplies of water required to meet various subarea water requirements as specified by the demand model. The basic inputs to the model are withdrawal, instream, and consumption requirements for the 50 NAR subareas and combinations thereof; data on existing and prospective intrabasin and interbasin transfers; and estimates of the costs of development. The formal objective of the model (the objective function) is to minimize the costs of supplying specified requirements. However, provision is made for inserting and varying parametrically constraints on sources, degrees, and types of development in the model, and in this way the influence of planning objectives other than cost minimization can be brought to bear on model results. Supply model runs are described and analyzed on T-331 to T-335.

The storage-yield relationships used in the supply model are based on analyses conducted with the aid of the third model. This hydrologic model is designed to determine the amount of storage required and the risk of failure of that storage at various demand rates at locations where the streamflow is known; and to express these failures by means of a shortage index. In the model monthly historic streamflows for groups of stations are extended to equal length, and then streamflow traces of 100 year length per series are generated. These streamflows are routed past a given station at predetermined storage increments to estimate the shortages obtained given preset yield requirements. Deficiencies are expressed in terms of an index computed by the model.

An interesting aspect of the NAR models for current planning operations is that all of them would now be suitable for implementation on advanced microcomputers (PC's). In this respect a replication or updating of the NAR methods would be far easier than were the original mainframe operations in a substantially less user-friendly environment.

2.3 INSTITUTIONAL ARRANGEMENTS

The main institutional measures developed and applied in the NAR study (Chapter 8) were shaped by the same context as the mathematical models: a framework plan; a commitment to multiobjective methods; large quantities of data from different agencies; and the use of many standard methods of economics, engineering, and water planning. These institutional methods included, first, the development of a planning structure specific to the NAR. This structure encompassed not only the prescribed Coordinating Committee but also an outside Board of Consultants and a Plan Formulation Work Group. Second, there was provision for substantial interaction and iteration throughout the planning process among the Coordinating Committee, the planning staff, and the staffs of the cooperating agencies (Major, 1977, pp. 56-57). Third, new reporting forms were developed for use by all participants throughout the planning process. These were designed to insure as far as possible that information would be focussed on the needs of the study rather than presented in terms of unrelated agency procedures. Fourth, the use of the models themselves was an institutional method that contributed to the coherence and cohesiveness of the study (Major, 1972). All of these methods were utilized within the three stage planning process described in Chapter 8.

2.4 NEEDS, SOURCES, AND DEVICES

This section describes the examination of interrelationships among alternative needs (outputs), sources, and devices in planning for the NAR. The examination of the interrelationships of these three elements occupied an important role in NAR planning from the beginning of the planning process. The analysis of these elements, often implicit in planning, was undertaken in an explicit and systematic way in the NAR study and with an awareness of the possible relationships of each of the three elements to the planning objectives (T-26, T-27, T-31 to T-33). The analysis stemmed from the planners' recognition (Report, 5) of the importance of moving away from an older water planning paradigm of requirements met by least cost alternatives to a broader, flexible examination in the light of objectives.

Fifteen water and water-related needs were examined in the NAR plan (Report, 123). These are shown in Table 2-1. The needs are outputs of water and related land systems, and are thus part of the physical relationship known as the production function, as are also the sources and devices. (When given weights [prices or multiobjective weights], needs, sources and devices enter into the objective function.) The modeling aspects of forecasting the needs are discussed in Chapter 5.

The sources that were examined in the NAR planning process include surface waters (rivers, lakes, fresh and saline estuaries, oceans); groundwater (fresh, mineralized, and saline); and atmospheric water (T-36; Report, ch. 4). Land resources (Report, ch. 4), discussed particularly in connection with visual and cultural needs, are described in Chapter 4 of this volume. Twenty-three major categories of devices were considered in the NAR (Report, 134-5). These are shown in Table 2-2.

In order to examine the possible interrelationships among needs, sources and devices systematically, a three dimensional matrix was developed to relate the three elements (T 35, T 36). In the matrix each cell represents one possible interrelationship among the needs, sources and devices. There are 4050 cells representing 15 needs, 9 sources, and 30 devices (as compared to the list of 23 devices used later in planning). An early analysis (T-27) concluded that there were 655 relationships of interest, or about 16% of the possible relationships. During the planning process, these were examined to focus on the most important relationships, and to eliminate from further systematic consideration the least important (T-27).

TABLE 2-1: NEEDS (Outputs)

- Publicly supplied water
- Industrial self-supplied water
- Rural water supply
- Irrigation water
- Power plant cooling
- Hydroelectric power generation
- Navigation
- Water recreation
- Fish and wildlife
- Water quality maintenance
- Flood damage reduction
- Drainage control
- Erosion control
- Health
- Visual and cultural environment

Source: Report, 123

TABLE 2-2: DEVICES

I. Resource Management

A. Water
- Storage facilities
- Withdrawal facilities
- Return facilities
- Conveyance facilities
- Quality control facilities
- Pumped storage facilities
- Desalting facilities
- Monitoring facilities

B. Water/Land
- Flood plain management
- Local flood protection
- Watershed management
- Erosion protection
- Drainage practices
- Waterway management

C. Land
- Land controls
- Land facilities

D. Biological
- Habitat management
- Fishways
- Fish Stocking

II. Research

III. Education

IV. Policy changes

A. Water Demand and allocation changes

B. Project operation changes

Source: Report, 134-135

2.5 REGIONAL SUBDIVISIONS FOR PLANNING

For planning purposes the NAR region was divided into geographical units at three different scales. The use of planning units is one of the many types of abstraction required to reduce real problems to a size that is manageable given the techniques current at the time of planning. The basic unit of planning for the NAR was the Area, of which there are 21 (Figure 2-1). Areas were defined primarily on the basis of hydrologic criteria. They vary in size from the largest, the Susquehanna, Area # 17, with 25,510 square miles, to the smallest, New York City, Long Island, and Coastal Westchester, # 13, with 1900 square miles. Some of the areas have several drainage basins; others only one, such as the Susquehanna; and still others have part of a drainage system (Report, 18; descriptions of the areas are in Report, 20-21). For some planning purposes the areas were aggregated into larger units, the six Subregions (A-F on Figure 2-1). For other planning purposes the areas were subdivided into Subareas (also called subbasins); these are shown as 21a, 21b etc. on Figure 2-1. These smaller units were used, for example, in the supply model. Economic and demographic data were generally available by county; for the purposes of using these data, area and subarea hydrologic boundaries were approximated by the relevant county boundaries. The counties included in each area and subarea are listed in T-211 to T-219. Plan results for the NAR were presented for the whole region (Report, ch. 8) and for the areas (Report, ch. 8 and Annex 1); as well as for the States (Annex 2).

2.6 PLANNING PERIOD AND FORECAST YEARS

The study plan period extended to the year 2020, with intermediate benchmark years of 1980 and 2000. The planning activities in the study and the final forecasts of needs, sources, and devices for different objectives took place in terms of these three forecast years. The study thus covered a period of approximately 50 years from the initially scheduled date of completion. The study was delivered to the Water Resources Council in 1972, so that the actual forecast time from date of delivery was 48 years.

The choice of benchmark years for the study can be thought of in the following way. The first benchmark year, 1980, provided a relatively near term target that was nonetheless sufficiently far off that some implementation activities could take place by that date if these were judged necessary. The second benchmark year, 2000, would provide for guidance in planning for, among other measures, large structural elements if these were recommended. The planning and implementation of these is a process that covers many years. The last benchmark year, 2020, would provide a distant marker for generalized long-range planning. This set of three benchmark years was judged sufficient to provide suitable planning guidelines at the framework level.

FIGURE 2-1 REGIONAL SUBDIVISIONS FOR PLANNING (Source: Report, 19)

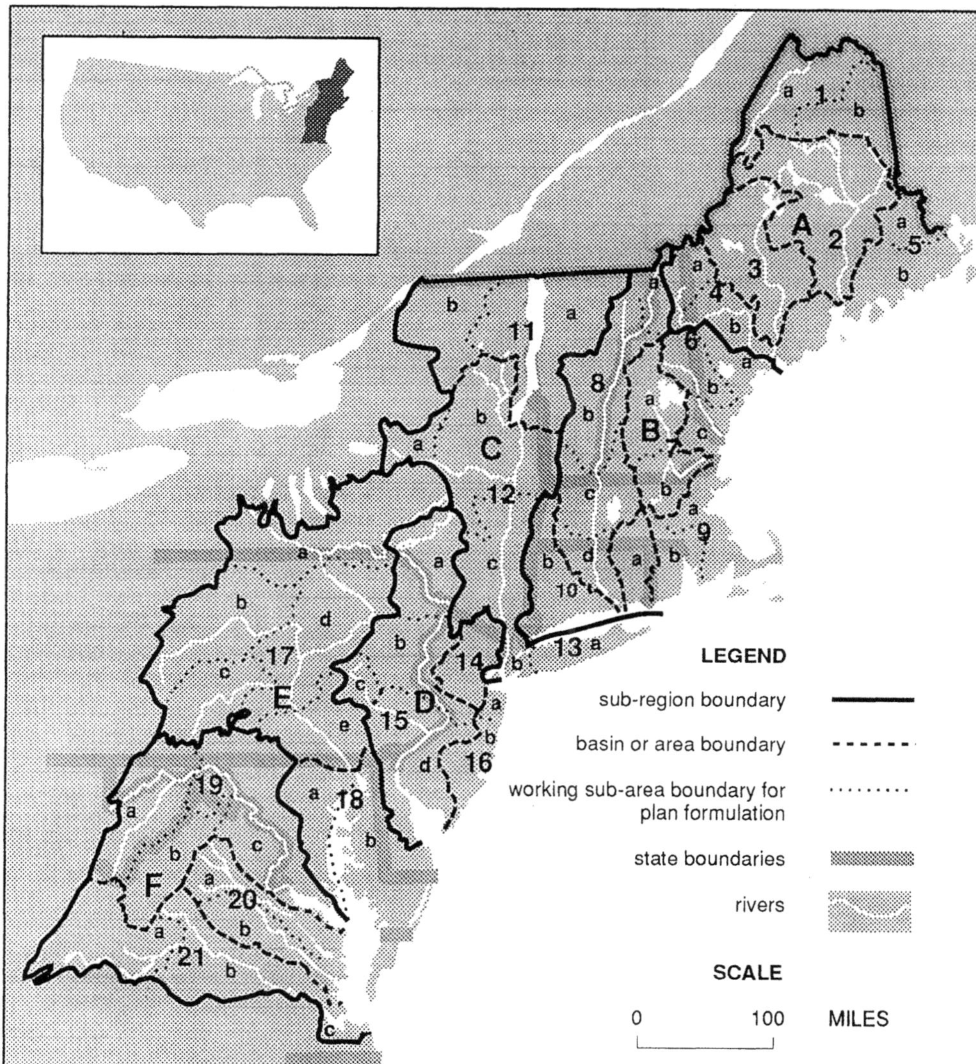

2.7 PERSPECTIVES ON THE METHODS

The six topics covered in this chapter are applicable to all major planning efforts. The first three topics, multiobjectives, mathematical models, and institutional arrangements, are discussed in later chapters. With respect to needs, sources, and devices, the appropriate definition of these will depend on specific regional conditions. The list of needs in the NAR study was based on an evaluation of conditions in the eastern United States; in other regions, some of the needs considered in the NAR study may not be of importance, while others should be added. For example, in some studies rural water supply will not be a relevant need; conversely, deep and shallow draft navigation will sometimes have to be treated as separate needs, as they were not in the NAR study. The definition of resources will also vary among studies, although here differences will probably not be so great as with needs. The selection of devices for a study will depend on topography and technology as well as on economic, political, legal, cultural and social factors including the effectiveness of institutions. For example, the difference in water law between the eastern and western United States affects the range of legal devices appropriate to water and related land planning in the two regions.

The definition of the planning region and its subdivisions depends on several factors, and the need for unequivocal definition of areas in practical planning work means that compromises among the factors will be required. Needs are normally related to economic and demographic variables, and data on these is typically organized according to administrative or political boundaries. Water and related land resources, on the other hand, are not usually well defined by these boundaries. For water resources, river basins or other hydrologic boundaries are appropriate; for land resources, physiographic and ecologic boundaries are relevant. In the NAR study the compromise adopted was to define areas and subareas by river basin or other hydrologic boundaries, and then to use for these areas economic and demographic data by county. This approach worked reasonably well in the NAR study because counties in the eastern United States are generally small relative to hydrologic boundaries and therefore provide fairly close approximations to them.

The benchmark years selected for a plan should be related to the nature and availability of economic and demographic forecasts relevant to the planning region, the type of devices that might be called for in the plan, and the lead times required for implementing these, including legal and administrative considerations. The use of three benchmark years, as in the NAR study, should be considered a minimal rule.

REFERENCES

Friedman, R., C. Ansell, S. Diamond, and Y. Y. Haimes, "The Use of Models for Water Resources Management, Planning, and Policy," Water Resources Research 20:7, 1984, 793-802.

Maass, Arthur, et al., Design of Water-Resource Systems, Harvard University Press, Cambridge, Mass., 1962.

Major, David C., "Impact of Systems Techniques on the Planning Process," Water Resources Research 8:3, June, 1972, 766-768.

Major, David C., Multiobjective Water Resource Planning, American Geophysical Union, Washington, D.C., 1977.

Major, David C. and Roberto L. Lenton, Applied Water Resource Systems Planning, Prentice-Hall, Englewood Cliffs, N.J., 1979.

Marglin, Stephen A., Public Investment Criteria, MIT Press, Cambridge, Mass., 1967.

Rogers, Peter P. and Myron B Fiering, "Use of Systems Analysis in Water Management," Water Resources Research 22:9, 1986, 146S-158S.

UNIDO (United Nations Industrial Development Organization), Guidelines for Project Evaluation, United Nations, N.Y., 1972.

PART II: APPLICATION OF THE NAR METHODS

Chapter 3: Multiobjectives

This chapter and Chapter 4 describe the use of multiobjective analysis in the NAR study. The first section of this chapter briefly describes the theory of multiobjective planning, and the following sections deal with: applications of this theory in NAR planning; the approach to plan formulation; the development of alternative multiobjective plans; and interest rates and other criteria. A final section provides perspectives on the methods. Because of its innovative quality, its extensiveness, and its importance in the study, the visual and cultural analysis used as part of the study's multiobjective approach is separately the subject of Chapter 4.

3.1 MULTIOBJECTIVE THEORY

Multiobjective analysis is a generalization of traditional benefit-cost analysis. The perceived need for this more general analysis in water resources planning arose from the observation that, while the objectives of water planning in the United States have historically been diverse, the traditional formal analytic criterion, benefit-cost analysis, restricted analysis to only one of many applicable objectives (Major, 1977, ch. 1). Detailed presentations of multiobjective theory can be found in Maass et al. (1962); Marglin (1967); UNIDO (1972); and Major (1977). Multiobjective standards for Federal water resources planning agencies are given in U.S. Water Resources Council (1973).

Multiobjective theory is shown graphically in Figure 3-1. This figure shows a multiobjective problem with two objectives, the traditional objective of increasing national income and increasing income to a particular region. (The rules for counting national and regional income differ; see the references given in the previous paragraph.) Benefits counted toward each objective are netted and discounted; that is, gross costs are subtracted from gross benefits in each time period of the analysis and the resulting net benefits are reduced to present value. The effects of various management and development options on the objectives are estimated and displayed in the net benefit space formed by the coordinates. These effects can be negative or positive for each objective. The group of attainable net benefit combinations is called the feasible set, and the boundary of this set is called the net benefit transformation curve. The points of interest to decision makers are the points on the boundary of the feasible set where it slopes from northwest to southeast. From any point on this part of the net benefit transformation curve no move can be made that is unambiguously good; a gain in net benefits toward one objective implies a loss of net benefits toward the other objective. Here preferences must come

FIGURE 3-1 MULTIOBJECTIVE THEORY (Source: Adapted from Major and Lenton, 1979, p. 31.)

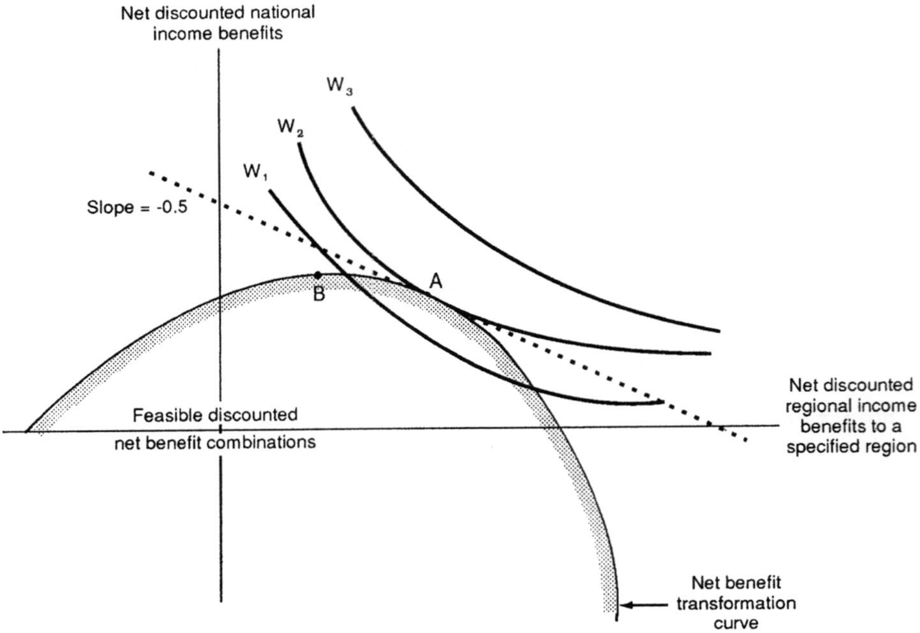

into play. These are represented by social welfare curves (W1, W2, W3 in Figure 3-1) Each social welfare curve is a locus of points in net benefit space of equal social utility; a curve further from the origin represents a higher level of social utility than a curve closer in. The formal purpose of multiobjective planning is to locate that point on the net benefit transformation curve that is tangent to the highest attainable social welfare curve. In Figure 3-1, this is point A, which shows the combination of net benefits toward the two objectives that is attained by implementing the optimal project or program. The negative of the slope of the tangent line at A gives the weight that society places on an additional dollar of net regional income benefit to the specified region in terms of an additional dollar of net income to the nation. In Figure 3-1, a slope of -0.5 means that, if the weight on an additional dollar of net national income is taken as 1.0 by convention, society places a weight of 0.5 on an additional dollar of regional income at the optimal point. (See Major, 1977, pp. 13-14, for details.) The traditional point of best design is shown by B, the highest attainable level of net discounted national income benefits. Although these results depend on the shapes of the net benefit transformation curve and the social welfare curves, it can be seen that in general A and B will not coincide. This in a nutshell is the reason for using multiobjective rather than traditional benefit-cost analysis in water planning. Details about the assumptions and qualifications of this analysis are in the references given earlier. It can be noted here that in practice planners do not have or need all of the information in Figure 3-1; in applications, the planner strives to develop that information that is necessary to the problem at hand. For example, preferences can be approached by discussion of alternative feasible points, rather than by direct attempts at estimating preference curves.

The representation in Figure 3-1 can be generalized to many dimensions. Beyond three dimensions, the presentation must be in numerical terms.

3.2 MULTIOBJECTIVES IN THE NAR

NAR planning took place in terms of three objectives: national income (the traditional objective of benefit-cost analysis); regional development; and environmental quality. These objectives and the multiobjective approach as a whole were adopted for NAR planning by the Coordinating Committee in September, 1967 (A-10), following preliminary discussions at the previous Committee meeting in April, 1967 (T-10). The staff paper that proposes and summarizes the reasons for this decision, "The Proposed Rationale for Plan Formulation," is reprinted in T-9 to T-15.

The three objectives selected by the Coordinating Committee were chosen for four reasons: first, they are objectives that might be significantly affected by water and related land resource development in the NAR; second, they are objectives that were of concern to the Coordinating Committee; third, they are objectives that were of broad interest to those dealing with a variety of public investment sectors in the United States at the time of planning; and fourth, they are reasonably comprehensive, capturing most of the effects of concern in water resources planning at the time of the NAR study (T-11).

The national income objective is defined in the standard benefit-cost manner (Marglin, 1967, pp. 40-44). National income in this sense refers to aggregate consumption, with valuation based on individual willingness to pay for system inputs and outputs, rather than to standard accounting definitions of national income. This objective is also sometimes referred to as the "economic" or "efficiency" objective. Its theoretical basis is a common model of welfare economics (Bergson, 1938; Graaff, 1963). The regional objective in the NAR is defined broadly to include regional income, output, employment, and other variables: it reflects a general idea of regional economic well-being. It is thus in principle an index objective, rather than one defined by a single metric such as jobs. The environmental objective as used in the NAR is also defined generally, to permit accounting for a variety of effects relating to the environment.

FIGURE 3-2 (Source: T-15)

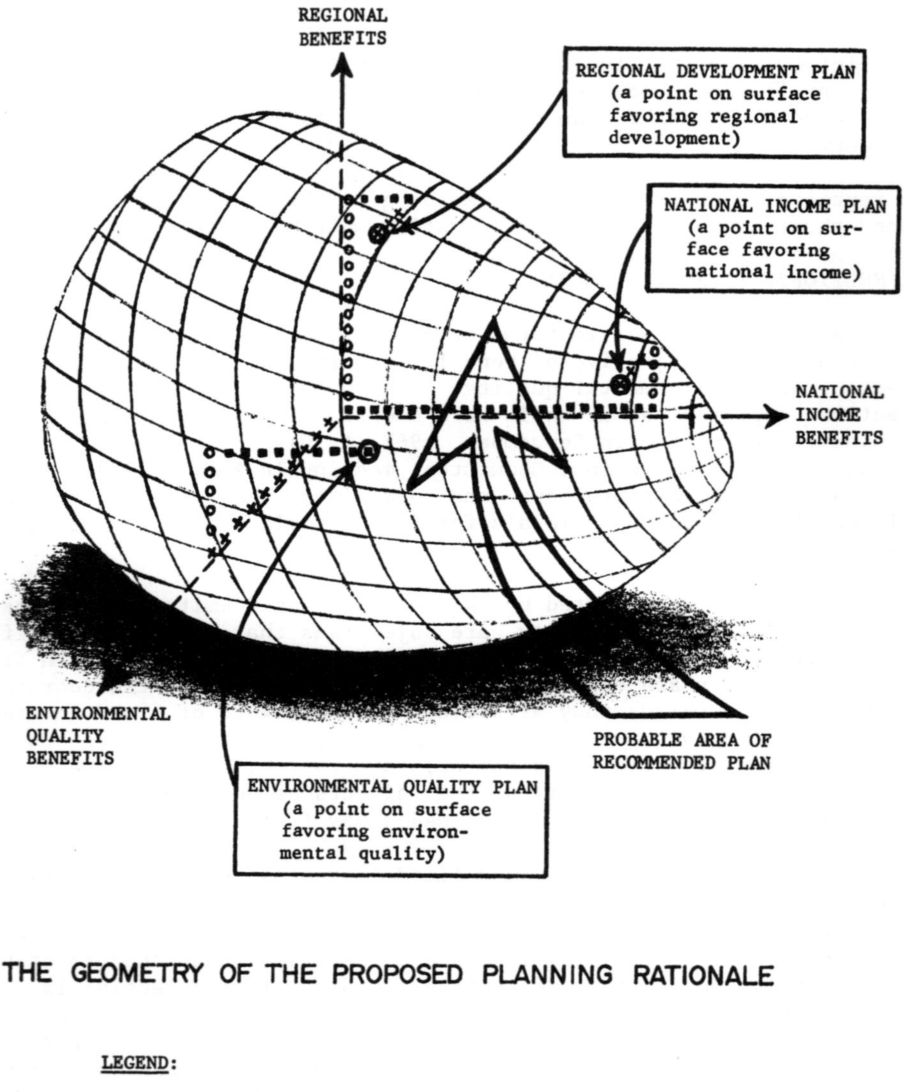

THE GEOMETRY OF THE PROPOSED PLANNING RATIONALE

LEGEND:

■■■■■■■■■■ National Income Components
oooooooooo Regional Development Components
xxxxxxxxxx Environmental Quality Components

Multiobjective planning as it was described graphically for the Coordinating Committee is shown in Figure 3-2. This figure displays a three dimensional transformation curve defined in terms of the three NAR objectives. The corresponding welfare curves (Figure 3-1) are not shown; rather, preferences were discussed in terms of the trade-offs implied by the choice of a point on the surface.

3.3 THE APPROACH TO PLAN FORMULATION

The approach to plan formulation for multiobjectives that was employed in the NAR study is shown in Figure 3-2. It was decided that the best approach was to proceed by first developing three alternative plans. Each of these would have a somewhat exaggerated emphasis on a single objective, and would thus represent a point on the transformation surface somewhat removed from the other two. (The plans were not designed to truly maximize with respect to single objectives because it was felt that such plans would not be relevant for final decision making.) These three plans would then be examined by the Coordinating Committee, which would use them to decide on a final recommended mixed objective plan for the NAR. The planners' goal was to show three alternative points on the surface that would, in effect, bound with reasonable closeness the area within which the final recommended plan would fall. Such a result is shown in Figure 3-2.

This approach to multiobjective plan formulation was carried out in a three stage planning process. This is summarized here and described in detail in Chapter 8 (see also Report, ch. 5). The first stage included: the adoption of multiobjective methods; the decision to develop three alternative draft plans, each emphasizing a particular objective; and independent estimates of needs, sources, and devices for each of the three alternative plans by the agencies responsible for studying individual project purposes. The second stage centered on the development of the demand and supply models and the improvement, integration, and expansion of the data and projections developed in the first stage. The result of the second stage included three draft alternative plans for each NAR area (T-81 to T-136), each responsive to a different objective. In the third stage the results of the demand and supply models were fully incorporated into plan formulation, and a draft mixed objective plan for each area was developed on the basis of the alternative plans. These draft mixed objective plans were the basis for final Coordinating Committee decisions on the recommended mixed objective plan for the NAR (see Report, ch. 8).

3.4 DEVELOPMENT OF ALTERNATIVE MULTIOBJECTIVE PLANS

The three alternative multiobjective plans for the areas and the NAR as a whole were developed by relating needs, sources, and devices to the objectives (T-32 to T-34). The three alternative plans thus each comprised a group of needs forecasts and projected sources and devices responsive to that plan's objective. Substantial efforts were made in the planning process to relate alternative forecasts of needs, and projected sources and devices to meet these, to the three planning objectives. The need estimates used for the three alternative plans are shown in Figure 3-3; the corresponding alternative devices, together with the relevant sources, are shown in Figure 3-4. The extent to which particular needs, sources, and devices were related to objectives in the planning process varied depending on planning resources (including agency capabilities) and the difficulty and significance of estimating the relationships involved. For example, many needs estimates responsive to different assumptions about objectives and parameters were made for the principal purposes incorporated in the demand model, including the three sets of estimates from the model shown in Figure 3-3. (A complete list of demand model runs requires six pages of Appendix T, T-274 to T-279.) By contrast, for some purposes few alternative estimates were made, and the estimates for each objective shown in Figure 3-3 are the same. Of the 15 purposes listed in Figure 3-3, needs forecasts for 9 show variations among objectives. Forecasts for 5 purposes do not vary, and no numerical forecasts were made for 1 purpose, health.

As examples, the methods used for forecasting needs for five purposes are described here. These are public water supply, industrial self-supply, and land drainage, for which forecasts vary with objectives, and flood damage reduction and water quality control, for which forecasts do not vary. In the case of flood damage reduction, devices rather then need levels vary with objectives. The approach used for forecasting for a sixth purpose, visual and cultural needs, is described in Chapter 4. Details of the methods used for each purpose are found in the respective appendices to the NAR study listed in the Introduction to this volume. The different methods used for various purposes were not always formally consistent; rather, the institutional methods described in Chapter 8 provided a framework for attempting to relate and to integrate them.

Public water supply and industrial self-supply needs forecasts were made within the demand model. Public water supply needs were forecast by means of a regression equation and industrial self-supply needs were forecast by means of the input-output component of the demand model (Chapter 5). To forecast needs for both purposes for the national income objective, the middle of three population and product per man hour growth rate forecasts from Appendix B, Economic Base (Section II, esp. B-8 and B-12) were used to generate the required population, per capita income, Gross National Product and regional output estimates for the demand model (T-222 to T-224). These rates are 1.3 % per annum for population and 3 % for productivity. For the environmental quality objective, it was assumed that population would be lower in the region than under national income assumptions reflecting, for example, greater emphasis on open space and land use planning. The lowest of the three Appendix B population growth rates, 1 % per annum, was used together with the middle productivity rate to generate environmental quality needs forecasts for public water supply and industrial self-supply (R-8). For regional development forecasts, it was assumed that population would grow at the middle rate of 1.3 % per annum, as for the national income forecast, but that productivity would grow at the highest of the three Appendix B rates, 3.2 %, reflecting regional development policies.

Land drainage needs were estimated outside of the demand model. These needs are for cropland requiring drainage and forest land which would be profitable to drain (J-14). (No wetlands drainage needs are forecast, for ecological reasons.) To forecast needs for the national income objective, it was assumed that then-current rates of land drainage would continue throughout the planning horizon. For the regional development forecasts, it was assumed that the treatment of both crop and forest land would be accelerated as compared to the national income assumptions. For example, cropland acreage drained by 1980, the first benchmark year, would be 150 % of base year acreage compared to 133 % for the national income forecasts. For environmental quality forecasts, it was held that cropland drainage needs would be the same as for the regional development objective on the assumption that open, working farmland is visually and culturally desirable (J-16). An increase in forest land drainage above the national income level, on the other hand, was believed to have both environmental advantages and environmental disadvantages; for this reason, forest drainage needs for the environmental quality objective were assumed to be the same as those for national income (J-16).

Two needs forecasts that do not vary by objective in Figure 3-3, the regional program, are also described here. Flood damage reduction needs and water quality control needs are both calculated outside of the demand model. Flood damages without mitigation measures are estimated on E-7 to E-12. These needs were taken to be the same for all three objectives. However, it was felt that the most appropriate flood damage reduction devices in an area might vary depending on the objective emphasized. The principal choice in this respect is between management and structural devices, with the former regarded as favoring the environmental quality objective. In six of the 21 NAR areas, the environmental quality flood damage reduction programs differ from the national income and regional development programs by an emphasis on management rather than structural devices (E-25). (An example of a substantial difference in programs on this account is Area 8, E-99, E-100.) In one area, Area 17, the national income and the environmental quality programs emphasize management measures as compared to the regional development program (E-151, E-152).

The water quality needs forecasts for the NAR, by contrast, are the same for all three objectives and also do not vary with respect to the recommended levels of waste treatment (Appendix L). This result came about both because of agency views on legal and other requirements for water quality control (Report, 214) and for organizational reasons

FIGURE 3-3 SUMMARY OF ALTERNATIVE LEVELS OF NEEDS
(Source: Report, pp. 124-125)

NEEDS-cumulative		Pres.	
Publicly Supplied Water	(1000 mgd)	5.5	
Industrial Self-Supplied Water	(1000 mgd)	3.9	
Rural Water Supply	(mgd)	400	
Irrigation Water: agriculture	(1000 afy)	200	
non-agriculture	(1000 afy)	100	
Power Plant Cooling: withdrawal, saline	(cfs)	23000	
brackish	(cfs)	12000	
fresh	(cfs)	10000	
consumption, brackish	(cfs)	120	
fresh	(cfs)	120	
Hydroelectric Power Generation	(1000 mw)	5	
Navigation: commercial	(m. tons annually)	600	
recreational boating	(m. boats)	1.6	
Water Recreation: visitor days	(m.)	(21)	
stream or river	(1000 miles)	(21)	
water surface	(m. acres)	(21)	
beach	(1000 acres)	(21)	
pool	(m. sq. ft.)	(21)	
land facilities	(1000 acres)	(21)	
Fish & Wildlife: sport fishing man-days	(m.)	100	
surface area, lake	(acres)	(12)	
stream	(acres)	(11)	
access, fresh	(1000 acres)	(21)	
salt	(1000 acres)	(11)	
anadromous	(acres)	(19)	
piers	(1000 feet)	(9)	
hunting man-days	(m.)	38	
access	(1000 sq. mi.)	(21)	
nature study man-days	(m.)	60	
access	(1000 acres)	(15)	
Water Quality Maint.: non-industrial	(m. PEs)	44	
industrial	(m. PEs)	70	
Flood Damage Reduction:			
avg. ann. damage, upstream	(m.$)	55	
mainstream	(m.$)	80	
tidal & hurricane	(m.$)	61	
Drainage Control: cropland	(m. acres)	1.2	
forest land	(1000 acres)	(0)	
wet land	(1000 acres)	(1)	
Erosion Control: agriculture	(m. acres)	15	
urban	(m. acres)	8	
stream bank	(1000 mi.)	(21)	
coastal shoreline	(1000 mi.)	(12)	
Health: vector control and pollution control		(21)	
Visual and Cultural			
landscape maintenance, unique natural	(sq.mi.)	11000	
unique shoreline	(mi.)	90	
high quality	(sq.mi.)	3800	
diversity	(sq.mi.)	(10)	
agriculture	(sq.mi.)	(7)	
landscape development, quality	(sq.mi.)	(6)	
diversity	(sq.mi.)	(1)	
metro. amenities	(mi.)	(1)	
met. amenities	(sq.mi.)	(12)	

FIGURE 3-3 (cont.)

ENVIRONMENTAL QUALITY			NATIONAL INCOME			REGIONAL DEVELOPMENT		
1980	2000	2020	1980	2000	2020	1980	2000	2020
6.9	9.6	13.1	7.2	10.6	15.7	7.2	10.8	16.1
7.0	11.8	17.9	7.0	12.4	20.0	7.2	13.2	22.2
←			570	790	720			→
1700	4700	5800	500	500	500	1700	4700	4700
300	490	720	270	460	720	280	470	720
43000	110000	190000	43000	117000	213000	43000	97000	170000
30000	52000	54000	30000	61000	97000	30000	66000	105000
11000	8000	4000	11000	50000	94000	11000	29000	48000
280	750	1160	280	540	860	280	660	1380
450	770	1280	450	1120	2340	450	1380	2750
←			15	42	100			→
700	900	1300	800	1100	1700	800	1200	2000
←			2.1	3.5	6.0			→
900	1500	2500	800	1300	2300	900	1500	2500
5.8	8.2	11.3	1.9	2.7	3.7	2.9	4.0	5.6
1.6	2.4	3.3	0.4	0.7	0.9	0.9	1.3	1.7
16	22	31	5	7	11	10	13	18
270	380	530	100	140	210	170	240	330
1020	1400	1990	180	260	370	370	500	690
←			110	140	180			→
←			75000	142000	248000			→
←			19000	32000	43000			→
←			2.0	4.5	7.4			→
←			5.4	15.2	26.8			→
←			890	1220	1590			→
←			140	400	700			→
←			43	53	66			→
←			5	18	29			→
←			70	88	109			→
←			150	410	730			→
←			56	70	86			→
←			140	300	620			→
←			82	145	275			→
←			130	260	530			→
←			96	181	359			→
1.8	2.7	2.9	1.6	2.4	2.5	1.8	2.7	2.9
0	170	670	0	170	670	170	670	2180
(0)	(0)	(0)	(1)	(0)	(0)	(0)	(0)	(0)
19	23	23	17	19	19	19	23	23
11	15	20	9	11	15	11	15	20
0.54	1.63	2.71	0.11	0.38	0.65	0.27	0.81	1.35
1.14	2.36	2.55	0.02	0.07	0.12	0.05	0.14	0.24
(21)	(21)	(21)	(21)	(21)	(21)	(21)	(21)	(21)
26000	26000	26000	26000	26000	26000	Same	as	EQ
1360	1360	1360	1240	1240	1240	Same	as	EQ
11200	18500	25900	11200	18500	25900	Same	as	EQ
3500	6900	10300	2800	5100	7400	Same	as	EQ
7300	7300	7300	7300	7300	7300	Same	as	EQ
1000	2000	3000	500	1000	1500	Same	as	EQ
300	300	300	150	150	150	Same	as	EQ
2	2	2	2	2	2	Same	as	EQ
670	670	670	410	670	670	Same	as	EQ

FIGURE 3-4 SUMMARY OF ALTERNATIVE LEVELS OF DEVICES
(Source: Report, 136-139)

DEVICES–incremental		Purposes	
I. Resource Management			
A. Water			
Storage Facilities ϕ			
reservoirs, upstream	(1000 af)	Irrig,Rec,FW,VC	
mainstream	(1000 af)	PS,FW,VC,Rec,WQM	
Withdrawal Facilities			
intakes & pumping, fresh	(mgd)	PS,Ind,Pow,Irrig	
brackish	(mgd)	Ind	
estuarine	(mgd)	Pow	
ocean	(mgd)	Pow	
wells	(mgd)	PS,Ind,Rur,Irrig	
Conveyance Facilities			
interbasin diversions, into	(mgd)	PS	
out of	(mgd)	PS	
Quality Control Facilities			
temperature, cooling towers & ponds		WQM,Pow,Rec	
chemical/biological			
potable water treat plants	(mgd)	PS	
waste treatment plants			
secondary (85%)	(m. PE removed)	WQM,VC,Rec	
secondary (90%)	(m. PE removed)	WQM,VC,Rec	
advanced (95%)	(m. PE removed)	WQM,VC,Rec	
effluent irrigation		WQM,VC,Rec,Irrig	
nutrient control		WQM,VC,Rec	
stormwater discharge control		WQM,VC,Rec	
acid mine drainage control		WQM,VC	
septic tank control		WQM,VC,Rec	
separate combined sewers		WQM,VC,Rec	
Pümped Storage		HPG	
Desalting Facilities			
Monitoring Facilities			
B. Water/Land			
Flood Plain Management			
upstream	(1000 acres)	FDR,VC,FW,Rec	
mainstream	(1000 acres)	FDR,VC,FW,Rec	
Local Flood Protection			
ocean	(projects)	FDR	
river	(projects)	FDR	
flood control channels	(mi.)	FDR	
Watershed Management	(1000 acres)	FDR,VC,Drn,FW,Rec	
Erosion Protection, land treatment		Ern	
coastal shoreline		Ern,Rec,VC	
river shoreline		Ern	
Drainage Practices		Drn,FW	
Waterway Management			
navigation channel improvement		Nav	
debris removal			
recreation boating facilities		Nav,Rec	

* From the supply model for the following purposes: PS, Ind, Rur, Irrig, Pow

ϕ Flood control storage not included.

FIGURE 3-4 (cont.)

ENVIRONMENTAL QUALITY			NATIONAL INCOME			REGIONAL DEVELOPMENT		
1980	2000	2020	1980	2000	2020	1980	2000	2020
1120	2180	790	210	80	40	1120	2180	40
610	1310	410	720	1710	940	750	1750	1030
2900	5000	6600	3000	5800	9100	3100	6400	10600
2600	3800	4800	2600	4400	6100	2700	4900	7200
(6)	(11)	(10)	(6)	(11)	(12)	(6)	(11)	(12)
(11)	(13)	(14)	(11)	(13)	(14)	(11)	(12)	(14)
1600	2600	2100	1300	2100	2100	1700	3000	2200
60	440	1090	80	690	1670	80	700	1830
60	440	1090	80	690	1670	80	700	1830
(13)	(14)	(18)	(10)	(12)	(16)	(11)	(15)	(20)
400	1700	2500	500	2200	3500	500	2200	3900
←			55	0	0			→
←			120	340	640			→
←			6	17	35			→
(1)	(1)	(1)	(1)	(1)	(1)	(1)	(1)	(1)
(20)	(20)	(20)	(8)	(8)	(8)	(10)	(10)	(10)
(19)	(15)	(15)	(9)	(5)	(5)	(10)	(6)	(6)
(3)	(1)	(0)	(3)	(1)	(0)	(3)	(1)	(0)
(9)	(9)	(9)	(5)	(5)	(5)	(6)	(6)	(6)
(19)	(14)	(14)	(8)	(4)	(4)	(9)	(5)	(5)
←			(7)	(11)	(13)			→
←			(0)	(0)	(0)			→
←			(0)	(0)	(0)			→
1260	3260	1570	250	50	170	210	60	170
(21)	(21)	(21)	(21)	(21)	(21)	(21)	(21)	(21)
11	3	0	11	7	0	11	7	0
180	190	130	130	190	90	210	180	70
580	330	300	840	1540	1850	2070	2140	80
8600	17100	17100	4100	4600	2900	6200	4300	1700
(21)	(21)	(21)	(21)	(21)	(21)	(21)	(21)	(21)
(13)	(13)	(13)	(10)	(11)	(11)	(11)	(11)	(11)
(21)	(21)	(21)	(18)	(21)	(21)	(21)	(21)	(21)
(20)	(20)	(20)	(20)	(20)	(20)	(20)	(20)	(20)
(1)	(0)	(0)	(13)	(14)	(8)	(13)	(14)	(8)
(0)	(0)	(0)	(0)	(0)	(0)	(0)	(0)	(0)
(21)	(21)	(21)	(21)	(21)	(21)	(21)	(21)	(21)

FIGURE 3-4 (cont.)

DEVICES–incremental (cont.)	Purposes	
C. Land		
Controls		
fee simple purchase (buying) (sq.mi.)	VC,Rec,FW	
fee simple purchase (buying) (mi.)	VC,Rec,FW	
purchase lease (sq.mi.)	VC,Rec,FW	
easements (sq.mi.)	VC,Rec,FW	
deed restrictions (sq.mi.)	VC,FW	
tax incentive subsidy (sq.mi.)	VC,FW	
zoning (sq.mi.)	VC,FW,Rec	
zoning (mi.)	VC,FW,Rec	
zoning and/or tax inc. subs. (sq.mi.)	VC,FW,Rec	
zoning and/or tax inc. subs. (mi.)	VC,FW	
Facilities		
recreation development	Rec	
overland transportation to facility	Rec	
parking and trails	FW,VC,Rec	
site sanitation and utilities	VC,Rec	
D. Biological		
Habitat Management, fish	FW	
wildlife	FW	
Fishways	FW	
Stocking, fish	FW	
wildlife	FW	
Water Quality Standards Enforcement	FW,WQM	
Insect Control	Hlth,Rec	
II. Research	WQM,Pow,Hlth,Rec	
III. Education		
IV. Policy Changes		
Water Demand and Allocation Changes		
pricing and rationing		
non-condenser power facilities	Pow	
re-circulation (internal)		
Project Operational Changes		
remove restrictions		
remove project	FW	
add new project needs	Rec,FW	
change project design load	Rec	
V. Others		
Upstream Flood Control Storage (1000 af)	FDR	
Mainstream Flood Control Storage (1000 af)	FDR	
Waste Water (mgd)	Ind	
Shell Fish Protection	Hlth	
Hydroelectric Generation Storage	HPG	
Septic Tank Elimination	Hlth	

FIGURE 3-4 (cont.)

ENVIRONMENTAL QUALITY			NATIONAL INCOME			REGIONAL DEVELOPMENT		
1980	2000	2020	1980	2000	2020	1980	2000	2020
20030	5130	5130	9340	80	50	Same	as	EQ
1210	0	0	780	0	0	Same	as	EQ
5400	0	0	2600	700	500	Same	as	EQ
4500	4500	4500	2200	1800	1800	Same	as	EQ
100	0	0	0	50	0	Same	as	EQ
0	0	0	550	550	550	Same	as	EQ
3800	0	0	5300	600	600	Same	as	EQ
60	0	0	350	0	0	Same	as	EQ
2400	2100	2100	14600	6400	6400	Same	as	EQ
0	0	0	32	0	0	Same	as	EQ
(21)	(21)	(21)	(21)	(21)	(21)	(21)	(21)	(21)
(15)	(15)	(15)	(13)	(13)	(13)	(16)	(16)	(16)
			(21)	(21)	(21)			
(21)	(19)	(19)	(21)	(20)	(19)	Same	as	EQ
←			(21)	(21)	(21)			→
←			(21)	(21)	(21)			→
←			(12)	(13)	(13)			→
←			(15)	(15)	(15)			→
←			(19)	(19)	(19)			→
←			(21)	(21)	(21)			→
←			(21)	(21)	(21)			→
(7)	(16)	(6)	(5)	(13)	(14)	(6)	(12)	(4)
(0)	(0)	(0)	(0)	(0)	(0)	(0)	(0)	(0)
(0)	(0)	(0)	(0)	(0)	(0)	(0)	(0)	(0)
(0)	(13)	(18)	(0)	(0)	(0)	(0)	(0)	(0)
(0)	(0)	(0)	(0)	(0)	(0)	(0)	(0)	(0)
(0)	(0)	(0)	(0)	(0)	(0)	(0)	(0)	(0)
←			(13)	(14)	(14)			→
(13)	(14)	(16)	(15)	(15)	(15)	(16)	(16)	(16)
(20)	(20)	(20)	(20)	(20)	(20)	(20)	(20)	(20)
1210	830	730	680	880	620	1060	840	540
500	390	5	620	1030	460	640	1030	460
140	310	600	140	340	720	150	380	820
←			(1)	(1)	(1)			→
(0)	(2)	(0)	(0)	(2)	(0)	(0)	(2)	(0)
←			(1)	(1)	(1)			→

discussed in Chapter 10. (The recommended levels of several devices relating to the water quality purpose in Figure 3-4, other than waste treatment plants, vary among the objectives.)

Devices and sources were related to objectives in the study in many ways. The most thoroughgoing effort to relate devices (and, implicitly, the corresponding sources) to objectives is in Appendix N, Visual and Cultural Environment, "Evaluation of the Visual, Cultural, and Ecological Impact of Devices" (N-124 to N-173); Chapter 4 in this volume. Others of the many examinations of the relationships of devices and sources to objectives include the analysis in the flood damage study described above, and the analysis of the relationship between the location and choice of energy facilities and objectives in the power analysis (P-104, P-109).

3.5 INTEREST RATES AND OTHER CRITERIA

Figure 3-1 shows contributions toward objectives as being both net (approached in the NAR by considering the positive and negative relationships of needs, sources, and devices to objectives) and discounted. Discounting requires the use of interest rates for each planning objective, to permit positive and negative effects toward that objective occurring in different points of time to be compared. Interest rates will, in general, vary as between objectives (Marglin, 1967, pp. 47-71; UNIDO, 1972, chs 13, 14; Major, 1977, pp. 36-41). In the NAR, an interest rate of 5 1/8 % (D-129, C-35) was used to discount the economic benefits and costs that were estimated. (In U.S. Federal water resource planning, the interest rate used for project evaluation is normally a weighted average of government borrowing rates; see the references just cited for critiques of this and similar approaches to determining the interest rate.) Interest rates for other objectives were not used explicitly. Rather, effects on other objectives were considered with reference to the time of their occurrence. Thus, in deciding among different alternative development and management plans for the NAR, decision makers would in effect be deciding implicitly on the interest rates (time weights) for these objectives (UNIDO, 1972, ch. 12).

Two additional investment criteria for project evaluation are budget constraints and optimal scheduling (Marglin, 1967, pp. 69-71 and 74-79; Major, 1977, pp. 42-47; Major, et al., 1987-88). The first integrates budgetary considerations explicitly into project design; the second optimizes the time of project implementation. These two criteria were not utilized in framework planning for the NAR; however, they will often be important for detailed basin and project planning.

3.6 PERSPECTIVES ON MULTIOBJECTIVES AND PLAN FORMULATION

The three objectives used in the study, national income, regional development, and environmental quality, should be applicable to many other studies. Nevertheless, objectives can be expected to differ as among studies, and they should therefore be chosen explicitly for every study, subject of course to guidance in national (or other applicable) policy (for example, U.S.Water Resources Council, 1973; see the discussion of this document in Major, 1977, pp. 5-6). The choice of objectives for a study is critical to the success of multiobjective planning: the use of too few objectives means that the planning problem will not be properly grasped by the planners, and the use of too many objectives introduces information that is not of substantial importance into an already complex decision problem. A useful criterion for choice of objectives is given in Major (1977, p. 9): "those objectives should be included in analysis that are important in terms of society's preferences and are those on which the range of measures under consideration is likely to have some significant effect." Marglin (1967, ch. 1), discusses the choice of objectives in the context of developing nations; see also UNIDO (1972).

The number of alternative plans developed during plan formulation can also be expected to vary among studies. The principle followed in the NAR study, to develop one alternative plan with a strong emphasis on each of the planning objectives, can be taken as a minimal rule. In many planning situations, the requirements of decision making will suggest additional plans representing different combinations of the planning objectives. The U.S. Water Resources Council's (1973) multiobjective Principles and Standards require two plans, one for each of the two design objectives (the same principle as in the NAR), and as many intermediate plans as are appropriate. (See Major, 1977, pp. 6, 16-18, for a discussion of the meaning of design objectives and display objectives as these terms are used in the Principles and Standards.) Note that the number of alternative plans discussed here does not include variations due to parameter sensitivity analysis. A substantial number of such analyses were performed in the NAR; see, for example, the run analysis for the demand model, T-271 to T-291.

REFERENCES

Bergson, A., "A Reformulation of Certain Aspects of Welfare Economics," Quarterly Journal of Economics 52:2, 1938, 310-334.

Graaff, J. de V., Theoretical Welfare Economics, Cambridge University Press, New York, 1963.

Maass, Arthur, et al., Design of Water-Resource Systems, Harvard University Press, Cambridge, Mass., 1962.

Major, David C., Multiobjective Water Resource Planning, American Geophysical Union, Washington, D.C., 1977.

Major, David C., Branden B. Johnson, Robert H. Cole, Richard Hosier, and Murdo Morrison, "Environmental Project Reevaluation with Rising Calendar Time Benefits," Journal of Environmental Systems 17:3, 1987-88, 167-175.

Major, David C. and Roberto L. Lenton, Applied Water Resource Systems Planning, Prentice-Hall, Englewood Cliffs, N.J., 1979.

Marglin, Stephen A., Public Investment Criteria, MIT Press, Cambridge, Mass., 1967.

UNIDO (United Nations Industrial Development Organization), Guidelines for Project Evaluation, United Nations, N.Y. 1972.

United States Water Resources Council, "Water and Related Land Resources: Establishment of Principles and Standards for Planning," Federal Register 38:174, 1973, 24778-24869.

Chapter 4: Multiobjectives: Visual and Cultural

This chapter describes the procedures used in the study for evaluating visual and cultural aspects of water and related land development and management. These procedures were important components of environmental quality planning and thus of multiobjective planning in the NAR, and they constitute one of the innovations of the study. A survey of landscape evaluation and an extensive bibliography on the subject are in Zube, Sell, and Taylor (1982).

The organization of the chapter is as follows. The first section describes the landscape quality evaluation method used in the study; the second section presents the methods used for evaluating the environmental effects of water planning and management devices; and the third section describes the development of visual and cultural needs estimates used in plan formulation. A final section provides perspectives on the methods.

4.1 LANDSCAPE QUALITY EVALUATION

This section describes the landscape quality evaluation method used in the NAR study. The method is based on four elements: a distinction between the underlying physical landform in an area and the superimposed human-influenced landscape system; a classification of landforms (called Landscape Series in the method) and superimposed human-influenced systems (called Landscape Units) into types; the ranking of each series and unit thus classified as high, medial, or low in quality; and the application of weighting factors to series and unit quality rankings to arrive at a composite index of landscape quality for each area in the NAR. This methodology is described in detail on N-16 to N-24; and the landscape ranking derived from it is shown for the NAR in the map in Figure 4-4. A general description of the purpose and scope of the visual and cultural work of which the landscape evaluation system is a part is given on N-1 to N-6.

The first element in the model is the concept that the regional landscape is composed of two major systems, the so-called natural and the human-made or the human-manipulated. Topography, vegetation and water are most often the elements associated with the visual image of the natural landscape. Of these, topography or landform is assumed to be the most dominant and the most enduring. This is the most nearly permanent component of the landscape and constitutes a basically fixed supply of both land area and landform. It is highly unlikely that any new mountain ranges or expanses of rolling hills will be created in the foreseeable future; for the next twenty or fifty years diastrophic erosive and depositional processes will have little impact on the visual character of regional landforms.

Superimposed upon this natural system is the visual manifestation of the presence or absence of humans. Humans alter the landscape in two basic ways: through the manipulation of natural resources in such activities as farming (cropland, pasture, orchards), mineral extraction and forest management (grid pattern planting and harvesting); and through the introduction of structures (buildings, roads, etc.). This human-made or human-manipulated landscape is one which is more susceptible to change than the landform component of the natural system. Within the human-manipulated system one can identify patterns which have a dominant image that is created as a result of the kind and extent of human manipulation. These patterns constitute a spectrum which ranges from landscape which appears to be completely human-made to that which is apparently untouched by humans. It includes landscape patterns which generate images of cities, towns, farms, forests and combinations of these.

The major pattern generating elements at the regional scale are open land (agricultural fields, pastures, wetlands) closed land (forests and woodlots) and human-made structures. The apparent presence or absence of these elements, their dominance in terms of percent of area covered and their distribution throughout the landscape are the primary determinants of the pattern image.

The NAR landscape quality evaluation system includes the identification and classification of both the landform (Landscape Series) and the landscape pattern (Landscape Units).

Landscape Series are divisions of the landscape which are identified by the general visual impression gained from the repetition of a dominant landform over a large area. The series classifications used in the NAR are: Mountain; Steep Hill; Rolling Hill; Undulating Land; Flat Land; and Coastline. A seventh classification, the Compound Series, encompasses those landscapes which are a product of two series such as parts of the Appalachian Range which alternate between Steep Hills and Rolling Hills. The series are illustrated in Figure 4-1.

Series are frequently but not always coincidental with established geomorphologic classifications. Landforms which are classified on the basis of their origin or their physical characteristics in youth (such as mountains) do not have the same visual characteristics in old age. For example, the southern peaks of the Green Mountains in Vermont and Massachusetts look like hills and do not fit the definition established for the Mountain Series. Therefore, while geomorphologic classifications and/or mapping of landscapes may frequently be coincidental with the series, they are not consistently so.

FIGURE 4-1 THE LANDSCAPE CONTINUUM (Source: N-19)

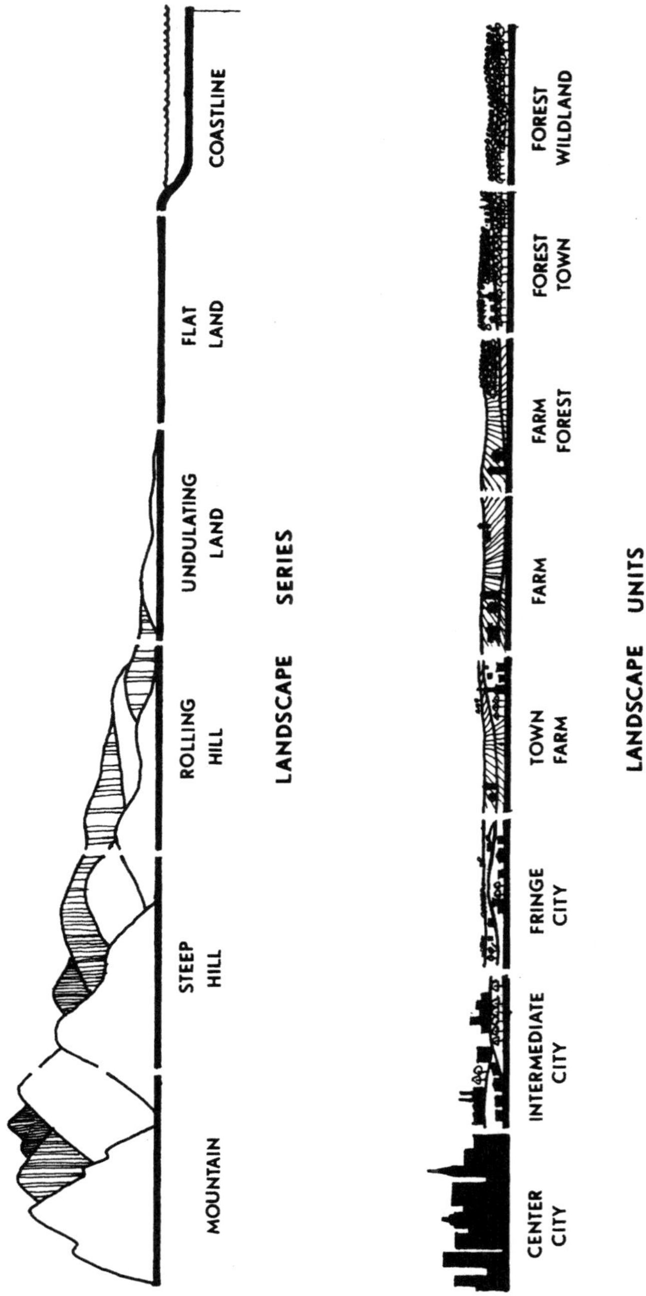

Landscape Series are divided into Sub-Series to identify differences in form and facilitate the process of qualitative evaluation (the details of Sub-Series classification and evaluation are given on N-50 to N-74 and N-93 to N-100). These differences are in three dimensional form and are generally the result of variations within a single landform component (e.g. long linear hill vs. short rounded hill, conical mountain vs. pyramidal mountain, linear coastline vs. irregular coastline) and/or variations in the distribution of the components (e.g. parallel rows of hills, scattered hills, clustered mountains). For example, within the Mountain Series the Adirondack Mountain Sub-Series is identified by densely clustered pyramidal peaks while the White Mountain Sub-Series (Figure 4-2) is identified by conical peaks which are randomly distributed in rows of varying lengths. Similar distinctions that help to define the visual characteristics of the landscape can be identified and described within most of the Landscape Series.

Landscape Units are identified by consistent ground pattern -- the two dimensional distribution of human-made structures and human-manipulated resources on the landscape. If the landscape is envisioned as a continuum ranging from that which is completely human-made to that which is apparently untouched by humans, Landscape Units represent cross sections of that continuum. The cross sections represent: Center City; Intermediate City; Fringe City; Town-Farm; Farm; Farm-Forest; Forest-Town; and Forest-Wildland. The units are illustrated in Figure 4-1 and the details of unit identification and evaluation are given on N-77 to N-89 and N-93 to N-100. These units are not separated by definitive boundaries but rather by zones of transition. The names are indicative of the dominant visual image imparted by the landscape so classified.

The major pattern generating elements in the Center City, Intermediate City and Fringe City are structures, open space and vegetation. Structures are viewed in reference to their size, distribution and spacing and their generalized use such as commercial, industrial or residential. Open space as a pattern generating element includes streets, parks and undeveloped land and is considered in terms of both scale or size and distribution or location. The presence or absence of vegetation, primarily trees, is a very important element in both the identification of city patterns and in their qualitative evaluation. Units identified with these elements may also be further qualified by the presence of visual misfits -- those factors which detract from the general visual image or impression. Misfits may be deteriorating structures, badly placed roads or freeways and the visual manifestations of polluted air and water. Entire districts in cities can become misfits because of their age and lack of continuous internal regeneration.

Human activity is also an important aspect of the urban and suburban landscape, particularly as reflecting the difference between activity in the Center City and in the intermediate and fringe areas. In the

FIGURE 4-2 WHITE MOUNTAIN SUB-SERIES (Source: N-52)

Conical peaks randomly distributed in short and long rows

WHITE MOUNTAINS **M 1**

NAR study it was assumed that the pattern generating elements of structure and open space provide a reasonable indicator of degrees of human activity commensurate with the scale of the study.

The Town-Farm, Farm, Farm-Forest, Forest-Town and Forest-Wildland Units are identified on the basis of population density, distance between towns and the amounts of open and closed land. Population density and distance between towns are the two factors which are used as a relative index of the distribution of human-made structures. Population density in all of these units except Forest-Wildland varies from 50 to 500 persons per square mile. In the Forest-Wildland Unit, density drops to less than 50 persons per square mile. In Town-Farm and Forest-Town Units, the distance between towns averages from two to five miles while in the other units there is a much wider range of distances separating towns. In the Forest-Wildland Unit, towns are on average more than five miles apart.

Open land and closed land are the two factors which are used as a general index of the apparent degree of resource manipulation. The percentages of land within the units in terms of farming or open land and forest or closed land varies and determines to a major degree the dominant unit image. For example, in the Farm Unit over 50 percent of the land is in open field agriculture whereas in the Farm-Forest Unit open field agriculture accounts for 20 to 35 percent of the land area. Farming or open field agriculture includes cropland and pasture land that may have scattered timber or shade trees but whose canopy covers less than 10 percent of the area. Forest also includes woodland which is in farm woodlots, except in the Forest-Town and Forest-Wildland Units where there is no significant amount of woodlot included in the forest or closed land percentage.

The preliminary classification of series and units was accomplished by interpretation of maps and population and land use data. Topographic maps (scale of 1:250,000) were used to delineate general landform areas. U.S. Bureau of the Census data were used for the initial delineation of general unit boundaries.

Travel cross sections of the NAR totaling over 10,000 miles were made on the ground and in the air. Initial field checks served to test the validity of the methodology; to identify the limitations of maps and statistical data for visual classification; and to see if there was a meaningful relationship between maps, the statistical data and actual visual recognition. Field checking also helped to refine the series definitions and pointed up certain consistent dimensional characteristics in the Mountain and Steep Hill Series. Field checking and making cross sections were equally important in developing the description of Landscape Sub-Series. The general identification of sub-series is possible using only map resources but an explanatory description of the visual characteristics requires actual inspection of the landscape. Topographic maps show relative dimensions and distribution but are inadequate for assessing the visual character of landform profile and specific form characteristics such as angular, rounded, linear or jagged. The landform analysis and description is most easily accomplished from the ground while pattern analysis and description is best done from the air.

The initial interpretation of the census data was directed at identifying differences within an individual Landscape Series. For example, these data indicated that within the Rolling Hills Series around Baltimore, Maryland and Washington, D.C. there would probably be two different units. To the north and northwest, more than 50 percent of the land was at the time of the analysis in open field agriculture. To the southwest only 20 to 35 percent of the land was in open field agriculture and 60 to 80 percent of the land was in forest and woodlots. In the former area the dominant image is that of a farm landscape. In the latter the image is a composite one of farms and forests. Initially these areas were simply identified as two unspecified units within the same sub-series. Similarly, units were identified in each series or sub-series without reference to units in other series or without reference to established pattern images. However, subsequent field checks and the analysis of field notes pointed up the recurrence of generalized visual images such as a farm landscape or a farm and forest landscape in widely separated parts of the NAR. Comparisons of the data for the areas identified as having similar images showed that there was a relationship between the separate units in the percent of land area devoted to open field agriculture (cropland and pasture) and to forest and woodland. There was also a general relationship in population density per square mile.

The landscape classification provides a means for obtaining a quantitative evaluation of different visual landscapes within the NAR. The relative abundance or scarcity of a given landscape can be assessed by areal measurement (or linear measurement for Coastline Series). Measurement can be limited to either series, sub-series or units or it can include composite evaluations such as the extent of a Farm-Forest Landscape Unit within a specific Rolling Hill Landscape Series. The quantitative evaluation can be related to the total region or to an individual river basin to assess the relative abundance or scarcity of each distinct landscape form and pattern within it.

Criteria were established for a qualitative evaluation of sub-series and units. Sub-series are analyzed on the basis of visual contrast and the diversity of spatial configuration created by landform. It is assumed that both three dimensional contrast (e.g. vertical mountain to horizontal valley) and spatial variety within the landscape are positive visual values. Spatial variety is judged on the basis of the shape of the spaces (simple or complex, linear or omni-directional), the degree of enclosure and the variety of shape and enclosure. Each sub-series is analyzed in reference to every other sub-series within a given series and assigned a value of high, medial or low. The procedure is followed for each Landscape Series in the region. The greater the degree of contrast and the greater the variety in spatial configuration, the higher the value.

Throughout most of the inland area of the NAR water is significant only as a site level feature. The visual impact of a water body is normally limited to its immediate environs. When 3% to 5% of the surface area of a unit is in water, however, and where there is fairly uniform distribution of the water bodies, the image of the landscape becomes water-oriented (Penobscot County, Maine: 4.4 %; Pike County, Pennsylvania: 3.1 %). When the percentage increases to approximately 5 to 10 percent with uniform distribution, water emerges as a dominant pattern element (Kennebec County, Maine: 9.0 %). These percentages are also relevant to the significance of water as an element of contrast in the series.

Landscape Units exclusive of the three city units were evaluated on the same three-level ranking of high, medial and low. The major criterion for the unit evaluation was variety or diversity within the pattern. Variety in landscape pattern is a result of the scale and distribution of: open land (crop fields and pastures); closed land (forest and woodlots); water surface (lakes, rivers and wetlands); and human-made structures (roads and buildings). The distribution of these elements on the land determines the richness or the monotony of the pattern. It is assumed that the more varied pattern has higher visual value; the three-level evaluation of high, medial and low is used. For example, a Forest-Wildland Unit consisting of uninterrupted forest is rated low. A similar area of forest interspersed with lakes, marshes and rivers would be rated medial or high because of its greater diversity (see Figure 4-3). The water surface not only contrasts with its surroundings as an element or material in the landscape but it also serves a function similar to that of open fields in creating pattern when it occurs in forest areas. Similarly a Farm-Forest Unit consisting of essentially uniform-sized blocks of pastures or hay fields alternating at regular intervals with forest areas or woodlots would not rate as highly as a pattern that was created by various sized blocks of farmland alternating at irregular intervals with various sized forest areas or woodlots. The consistent presence of water affects ratings not only as an element of strong contrast within the landscape but also because it influences the two dimensional pattern.

Detailed inventories and evaluations of the three city Landscape Units were not made both because of the regional thrust of the NAR study and because of the total resources that would have been required for these tasks (N-78). General inventory and evaluation perspectives for the city units are given on N-78 to N-84 and N-90; the importance of the cities in the NAR study is made evident throughout Appendix N.

FIGURE 4-3 (Source: N-89)

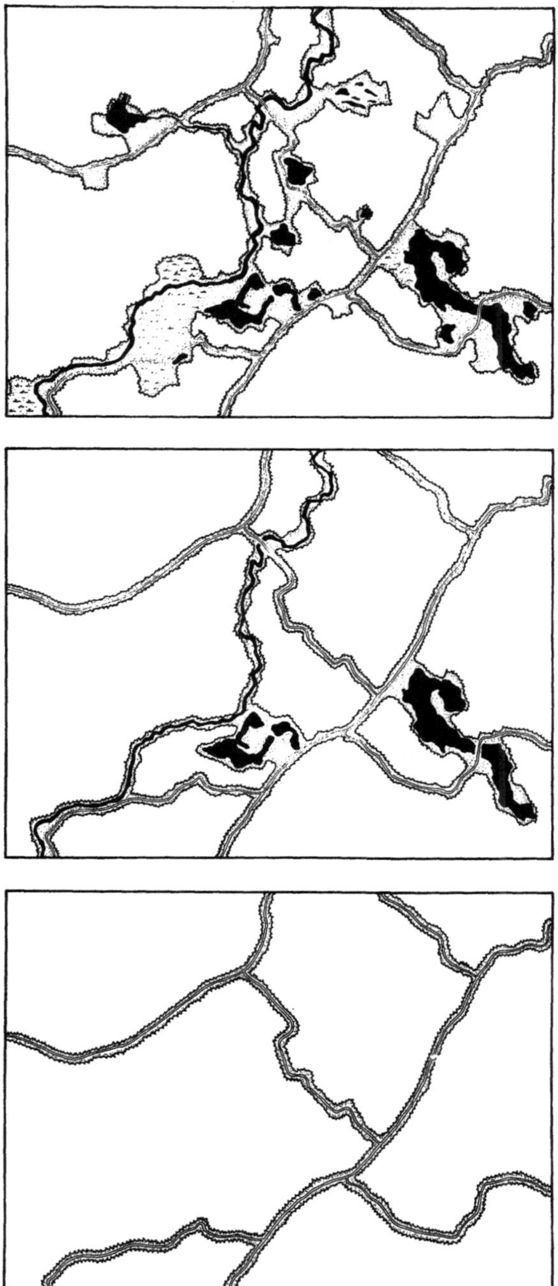

A numerical indicator of the combined landscape evaluation (Series - landform, Unit - pattern) is obtained by assigning numerical values to the high, medial and low evaluations (9, 6 and 3 respectively) and by assigning weighted values to series or sub-series and units. Weighting values are based on the hypothesis that the more dominant the form (Series) the less important the pattern (Unit) for high visual value. For example, pattern is not important for high value in mountains but it is very important in flat land. In the case of the latter, the pattern is the major variety or diversity generating element. (The Coastline Series was evaluated solely on the basis of the series analysis, and the three City Units were not evaluated with this method.) Weighting values for series and units are:

Series	Series Weighting (SW) Value	Unit Weighting (UW) Value
MOUNTAINS	9	1
STEEP HILLS	7	3
ROLLING HILLS	5	5
UNDULATING LAND	3	7
FLAT LAND	1	9

The combined visual landscape value equals:

CLV = (SV x SW) + (UV x UW)

Where: CLV = Combined landscape value
SV = Series evaluation (high 9, medial 6, low 3)
SW = Series weighting value
UV = Unit evaluation (high 9, medial 6, low 3)
UW = Unit weighting value

Possible combined scores range from a low of thirty to a high of ninety. Combined evaluations are ranked as follows: a score of 70 to 90 is high; a score of 50 to 69 is medial and a score of 30 to 49 is low. The landscape rankings for the NAR resulting from the application of this method are shown in the map in Figure 4-4.

4.2 ENVIRONMENTAL EFFECTS OF WATER PLANNING AND MANAGEMENT DEVICES

This section describes the approach used in the study for evaluating the impacts on the environmental objective of different water resource planning and management devices. (The analysis is given in full on N-118 to N-173.) The list of devices used in NAR planning is first analyzed in terms of the extent to which each is likely to be used for satisfying visual and cultural needs. The results of this analysis are given in Figure 4-5. Among the dozen most important devices are six legal devices; the relevance of these to the maintenance and improvement of visual and cultural quality is analyzed on N-120 to N-123. Next, each physical device, together with three program measures that are partly physical (flood plain management, watershed management, and preservation), is analyzed in terms of its potential visual, cultural, and ecological impacts. Summaries of these analyses are provided in

FIGURE 4-4 LANDSCAPE QUALITY RANKING IN THE NAR
(Source: Endleaf Map Figure N-38)

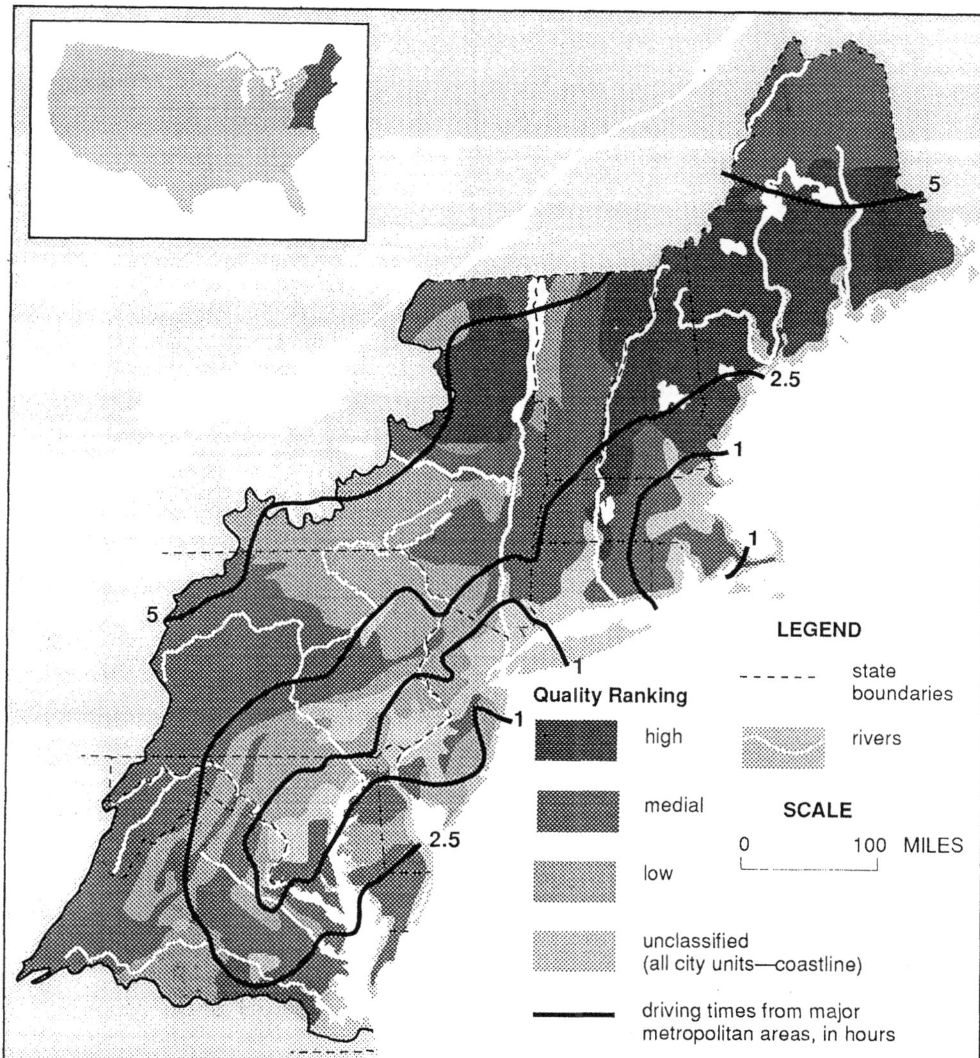

FIGURE 4-5 DEVICES MOST LIKELY TO BE USED FOR SPECIFIED VISUAL AND CULTURAL NEEDS (MARKED WITH AN X) (Source: N-119)

DEVICES	Preserve Unique Natural & Cultural Landscapes	Protect Quality Landscapes	Protect Composite Landscapes	Develop Quality Landscapes	Develop Clean Water	Develop Metropolitan Amenities
Legal: Fee Simple Purchase	x					x
Purchase-Lease Back	x		x			x
Easements	x	x	x	x		x
Deed Restriction	x	x	x			
Zoning	x	x	x		x	x
Tax Incentive and/or Subsidy	x		x	x	x	x
Demand Constraint		x	x			
Education						
Diversion						
Impoundment				x		x
Ground Water Management Facility						
Weather Modification						
Channel Improvement						
Clearing & Cleaning				x		
Local Flood Protection						
Bank Protection						
Coastal Protection						
Water Supply Installation				x		x
Water Renovation Installation						
Desalination Installation						
Off-Stream Cooling Installation						
Hydroelectric Power Installation						
Navigation Facility						
Recreation Facility				x		x
Fish & Wildlife Facility		x	x	x		x
Drainage Facility						
Waste Disposal Facility		x		x		
Flood Plain Management		x	x	x	x	
Watershed Management		x	x	x	x	
Preservation	x	x				
Research						
Waste Treatment		x		x	x	

tabular form that show the general level of visual, cultural, and ecological impacts of each device on each Landscape Series and Landscape Unit. The tables for the visual impacts of devices on series and units are given here as Figures 4-6 and 4-7. An example of the detailed analysis of a single device (dikes, levees, and flood walls) is shown in Figure 4-8. A similar analysis was done for each of the physical and program devices; these were designed for use in evaluating the impacts on the environmental objective of alternative draft plans for the NAR.

4.3 VISUAL AND CULTURAL NEEDS

As part of the NAR approach of developing separate draft plans emphasizing different objectives, visual and cultural needs for the region were estimated. These needs were developed by considering six categories of visual and cultural concerns: the preservation of unique natural and cultural landscapes; the protection of landscape quality; the protection of composite landscapes; the development of landscape quality; the development of improved water quality; and the development of urban amenities. Each of these was analyzed in terms of the NAR as a whole with reference to specific areas or Landscape Series and Units (N-103 to N-114); then the needs were presented by each of the 21 areas in the NAR by benchmark year (N-174 to N-274). It was assumed for visual and cultural needs (this is a special assumption) that regional aspirations in the NAR would include concern for a high level of environmental preservation and improvement, so that the same forecasts of visual and cultural needs were used for both the EQ and RD objectives (N-36). The forecast for the NI objective generally suggests lower concern for and investment in environmental preservation and improvement than the EQ/RD forecast. The two forecasts, EQ/RD and NI, thus differed in terms of needs foreseen, investments to meet these, and the devices chosen. An example basin plan for visual and cultural needs is that for Basin 10, reproduced here as Figure 4-9. These basin plans were used in the formulation of draft alternative plans and ultimately the final recommended NAR plan.

4.4 PERSPECTIVES ON VISUAL AND CULTURAL EVALUATION

The methods described in this chapter were important components of environmental quality planning, and thus of multiobjective planning, in the NAR study. These methods help to illuminate and thus ultimately resolve the complex decisions on multiobjectives involved in water resources planning. There are many examples of water resources projects and programs where planners have not used such methods and as a result have not provided decision makers with an adequate basis for choice. For instance, the effects of a proposed dam on the Danube at Nagymaros in Hungary on a unique and historic landscape were neglected in planning, and became a major reason for opposition to construction of the proposed dam. (This project is described in Carbonell and Yaro, et al., 1989.) The systematic consideration, early in the planning

FIGURE 4-6 POTENTIAL VISUAL IMPACT OF WATER MANAGEMENT DEVICES ON LANDSCAPE SERIES (Source: N-126)

PHYSICAL DEVICES	M	SH	RH	UL	FL	C	X
Filling and Excavation	NNA	NNA	Lp-Ln	Lp-Ln	Lp-Ln	NNA	Lp-Ln
Clearing & Cleaning	NNA	NNA	Lp-Ln	Lp-Ln	Lp-Ln	NNA	Lp-Ln
Bank Protection	NNA	NNA	Lp-Ln	Lp-Ln	Lp-Ln	NNA	Lp-Ln
Diversion	NNA	NNA	NNA	O-Hn	O-Hn	NNA	O-Hn
Dikes, Levees, Flood Walls	NNA	NNA	Ln-Hn	Ln-Hn	Ln-Hn	NNA	Ln-Hn
Sea Walls, Breakwaters, Sea Levees, etc.	NNA	NNA	NNA	NNA	NNA	Lp-Hn	NNA
Sand Fill-Beach Stabilization	NNA	NNA	NNA	NNA	NNA	Hp-Lp	NNA
Power Generating Installation	O-Ln	O-Ln	Ln	Ln-Hn	Ln-Hn	Ln-Hn	Ln-Hn
Waste Treatment & Water Renovation Inst.	NNA	NNA	Lp-Ln	Lp-Ln	Lp-Ln	NNA	NNA
Desalination Installation	NNA	NNA	NNA	NNA	NNA	O-Ln	NNA
Off Stream Cooling Installation	O-Hn	O-Hn	O-Hn	O-Hn	O-Hn	Ln-Hn	Ln-Hn
Dams & Hydroelectric Power Installations	Lp-Ln	Lp-Ln	O-Ln	O-Hn	Ln-Hn	Ln-Hn	Lp-Ln
Recreation Facility	Lp-O	Lp-O	Lp-O	Lp-O	Lp-O	Hp-Ln	Lp-O
Fish & Wildlife Fac.	Lp-O	Lp-O	Hp-O	Hp-O	Hp-O	Hp	Hp-O
Water Supply Facility	Lp-O	Lp-O	O-Ln	O-Ln	O-Ln	O-Ln	Lp-O
Ground Water Management Facility	NNA	NNA	Lp-Ln	Lp-Ln	Lp-Ln	Lp-Ln	Lp-Ln
Drainage Facility	NNA	NNA	Lp-Ln	Lp-Ln	Lp-Ln	Lp-Ln	NNA
Navigation Facility	NNA	NNA	NNA	Lp-Ln	Lp-Ln	Lp-Ln	NNA
Waste Disposal Fac.	NNA	NNA	Ln-Hn	Ln-Hn	Ln-Hn	Ln-Hn	Ln-Hn
Impoundments	Hp-Lp	Hp-Lp	Hp-Lp	Hp-Lp	Hp-Lp	NNA	Hp-Lp
PHYSICAL/NON-PHYSICAL DEVICES							
Flood Plain Mgmt.	NNA	NNA	NNA	Hp-Hn	Hp-Hn	Hp-Hn	Hp-Hn
Watershed Mgmt.	Hp-O	Hp-O	Hp-O	Hp-O	NNA	NNA	Hp-O
Preservation	Hp-O	Hp-O	Hp-O	Hp-O	Hp-O	Hp-O	Hp-O

LEGEND

NNA	Not normally applicable	M	Mountains
Hp	Moderate to high positive visual impact	SH	Steep Hills
Lp	Low to moderate positive visual impact	RH	Rolling Hills
O	No significant change in visual impact	UL	Undulating Land
Ln	Low to moderate negative visual impact	FL	Flat Land
Hn	Moderate to high negative visual impact	C	Coastline
		X	Compound

FIGURE 4-7 POTENTIAL VISUAL IMPACT OF WATER MANAGEMENT DEVICES ON LANDSCAPE UNITS (Source: N-127)

PHYSICAL DEVICES	CC	IC	FC	TFa	Fa	FaFo	FoT	FoW
Filling and excavation	Lp-Ln	Lp-Ln	Lp-Ln	Lp-Ln	Lp-Ln	Lp-Ln	Lp-Ln	Lp-Ln
Clearing & Cleaning	Lp-Ln	Lp-Ln	Lp-Ln	Lp-Ln	Lp-Ln	Lp-Ln	Lp-Ln	Lp-Ln
Bank Protection	Lp-Hn	Lp-Hn	Lp-Ln	Lp-Ln	Lp-Ln	Lp-Ln	Lp-Ln	Lp-Ln
Diversion	Lp-Hn	Lp-Hn	Lp-Hn	O-Hn	O-Hn	O-Hn	O-Hn	O-Hn
Dikes, Levees, Flood Walls	Ln-Hn	Ln-Hn	Ln-Hn	Ln-Hn	Ln-Hn	Ln-Hn	Ln-Hn	Ln-Hn
Sea Walls, Breakwaters, Sea Levees, etc.	NNA	Lp-Ln	Lp-Ln	NNA	NNA	NNA	NNA	NNA
Sand Fill-Beach Stabilization	NNA	Hp-Lp	Hp-Lp	NNA	NNA	NNA	NNA	NNA
Power Generating Installation	Ln-Hn	Ln-Hn	Ln-Hn	O-Hn	Ln-Hn	Ln-Hn	Ln-Hn	O-Hn
Waste Treatment & Water Renovation Ins.	NNA	Lp-Ln	Lp-Ln	Lp-Ln	NNA	NNA	NNA	NNA
Desalination Installation	NNA	Lp-Ln	O-Ln	NNA	NNA	NNA	NNA	NNA
Off Stream Cooling Installation	Ln-Hn	Ln-Hn	Ln-Hn	O-Hn	O-Hn	O-Hn	O-Hn	O-Hn
Dams & Hydroelectric Power Installation	NNA	NNA	O-Hn	Lp-Hn	O-Hn	Lp-Hn	Lp-Hn	Lp-Hn
Recreation Facility	Hp-Lp	Hp-Lp	Lp-O	Lp-O	Lp-O	Lp-O	Lp-O	Lp-O
Fish & Wildlife Fac.	NNA	NNA	NNA	Hp-O	Hp-O	Hp-O	Hp-O	Hp-O
Water Supply Fac.	Lp-Ln	Lp-Ln	Lp-Ln	Lp-Ln	O-Ln	O-Ln	O-Ln	Lp-Ln
Ground Water Management Facility	Lp-Ln	Lp-Ln	Lp-Ln	Lp-Ln	Lp-Ln	Lp-Ln	Lp-Ln	Lp-Ln
Drainage Facility	NNA	NNA	NNA	Lp-Ln	Lp-Ln	Lp-Ln	NNA	NNA
Navigation Facility	Lp-Ln	Lp-Ln	Lp-Ln	Lp-Ln	Lp-Ln	Lp-Ln	Lp-Ln	NNA
Waste Disposal Fac.	Ln-Hn	Ln-Hn	Ln-Hn	Ln-Hn	NNA	NNA	Ln-Hn	NNA
Impoundments	NNA	NNA	Hp-Lp	Hp-Lp	Hp-Lp	Hp-Lp	Hp-Lp	Hp-Lp
PHYSICAL-NON-PHYSICAL DEVICES								
Flood Plain Mgmt.	Hp-Hn	Hp-Hn	Hp-Hn	Hp-Ln	Hp-Hn	Hp-Hn	Hp-Hn	Hp-Hn
Watershed Mgmt.	NNA	NNA	NNA	Hp-O	Hp-O	Hp-O	Hp-O	Hp-O
Preservation	Hp-O	Hp-O	Hp-O	Hp-O	Hp-O	Hp-O	Hp-O	Hp-O

LEGEND

NNA	Not normally applicable
Hp	Moderate to high positive visual impact
Lp	Low to moderate positive visual impact
O	No significant change in visual impact
Ln	Low to moderate negative visual impact
Hn	Moderate to high negative visual impact

CC	Center City
IC	Intermediate City
FC	Fringe City
TFa	Town-Farm
Fa	Farm
FaFo	Farm-Forest
FoT	Forest-Town
FoW	Forest-Wildland

FIGURE 4-8 ANALYSIS OF A SINGLE DEVICE (Source: N-144 to N-146)

Dikes, Levees and Flood Walls: These devices are of concrete, metal or earth and are generally under twenty feet in height and exposed on one or both sides. They are built upon dry ground, generally parallel to a water body, to confine and control flood water. They separate the river from the developed areas and provide real and psychological security for the inhabitants of the flood plains.

The devices are normally long and continuous. They have uniform height and alignment and are often taller than the average person. They make minimal space demands upon the landscape but have great visual, cultural and ecological impact. The height, length and material are important elements in determining the degree of visual, cultural or ecological impact.

The devices can be used alone or in combination with each other. They are most often used in Rolling Hill, Undulating Land and Flat Land Series.

Visual Impact

In areas with higher topographic variation and diversity in pattern the variation and diversity tend to interrupt the linear quality making the particular device less monotonous and dominant.

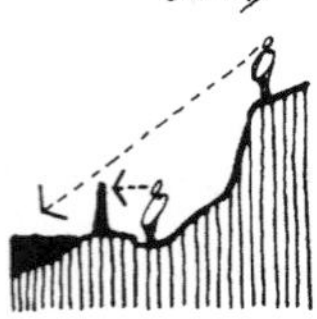

Topographic variation increases the possibility of regaining visual contact with water from points of varying elevation which are above the top elevation of the device.

At lower elevations the view of the water is cut off along most of its length.

Their form and location is frequently unrelated to adjacent natural or man-made features.

Dikes and levees, made of natural materials, contrast less with their natural surroundings than do flood walls which are of man-made materials (except in City Units where man-made materials predominate).

FIGURE 4-8 (cont.)

From the design standpoint, dikes and levees are more flexible than flood walls and thus are more likely to have a satisfactory design relationship with their surroundings. For example, the concept of levees can be expanded to a series of man-made hills, the tops of which would be free from flooding and could be used for building sites and the valleys or flood-prone lands could be used for parking, roads and open space. Visual and physical access to the water would thus be maintained and potentially enhanced.

Cultural Impact

By providing security they increase the potential for intensity of land use development and investment.

By cutting off access to the water, opportunities for recreational use are limited. Dikes and levees provide greater opportunity for additional uses such as recreation than do flood walls because they make access to the water easier.

Flood hazards for areas located both above and below the project may develop, giving rise to the need for additional management and protective measures.

Ecological Impact

These devices make a very small impact on the ecology of the stream but they do have an impact on the terrestrial ecology of the flood plain, not only on the site but above and below the site as well.

The deposition of silt on the flood plains is reduced, consequently reducing a natural source of fertility for flood plain soils.

By straightening and deepening the flood channel, water flows more rapidly and carries its silt load farther downstream to be deposited on the flood plain or on the deltas at the mouth of the stream. Increased deposition downstream will alter the make-up of the aquatic plant community by changing the channel substrata or cover over shellfish beds or fish spawning areas.

These devices may increase silt deposition upstream from the device in both the normal flow channel and the flood plain. This reduces spawning areas and promotes rooted aquatic vegetation in the slower moving water upstream.

FIGURE 4-9 EXAMPLE OF A BASIN PLAN (Source: N-214,N-215)

BASIN 10

Approximately four-fifths of the basin is composed of Rolling Hills and one-fifth is of Steep Hills. In this relatively urban basin with one-fourth of the area in City Units, there is a predominant pattern of Forest-Town landscape (68%) and a small amount (7%) of Town-Forest Units. Overall landscape quality for this basin is medial.

Population density averages 506 persons per square mile (1960).

The Rolling Hill, Forest-Town landscape of high quality found on the New York-Connecticut line is within a short driving time of major metropolitan areas and this warrants protection from the expanding urban pressures. The coastline in the basin is already extensively developed but access for public recreation purposes is needed. There is a clustering (5) of historic sites in the New Haven vicinity. Needs include the preservation of shoreline, development of quality landscapes, protection of landscape diversity, provision of clean water for the entire area (100%) and the provision of urban amenities.

LANDSCAPE INVENTORY

SERIES	AREA	UNITS	AREA
Steep Hills	800 sq.mi.	City	1100 sq.mi.
Rolling Hills	3500 sq.mi.	Town-Farm	300 sq.mi.
		Forest-Town	2900 sq.mi.

NEEDS, DEVICES AND COSTS - BASIN 10

	Environmental Quality			National Efficiency		
	1980	2000	2020	1980	2000	2020
NEEDS						
Preserve Coastline (linear mi.)	60			70		
Protect Coastline (linear mi.)	20			10		
Protect Landscape Diversity (sq.mi.)	30	30	30	15	15	15
Develop Landscape Quality (sq.mi.)	300	300	300	150	150	150
Develop Clean Water (% of basin)	100%			100%		
Develop Metropolitan Amenities (sq.mi.)	55			30	25	
LEGAL DEVICES						
Fee Simple (sq.mi.)	370	315	315			
Fee Simple (linear mi.)	20			10		
Purchase-Lease Back (sq.mi.)				180	175	150
Easements (sq.mi.)	15	15	15	15	15	15
Zoning (linear mi.)	60			70		
OTHER DEVICES						
Impoundment	x	x	x	x	x	x
Water Supply	x	x	x	x	x	x
Waste Treatment	x			x		
Recreation Facility	x	x	x	x	x	x
Wildlife Facility	x	x	x	x	x	x
Preventive Flood Plain Management	x	x	x	x	x	x
Corrective Flood Plain Management	x	x	x	x	x	x
Watershed Mgmt., Agricultural Practices	x	x	x	x	x	x
Watershed Mgmt., Reforestation	x	x	x	x	x	x
COST IN $ MILLION						
First Cost	395.0	98.5	98.5	198.0	56.5	49.0
Annual Return				1.0	1.0	1.0

process, of landscape values as they are perceived by the people affected in the region might well have resulted in a different project and avoided a bitter controversy.

The development of social objectives in recent decades suggests that visual and cultural assessments should be part of every water resources plan, including those for developing regions. Since the completion of the NAR plan many methods have become available for landscape, and more broadly, visual and cultural evaluation (Zube, Brush, and Fabos, 1975; Smardon, Palmer and Felleman, 1986). The methods chosen for a particular study will depend on the nature of the visual and cultural resources to be considered, the objectives that are meaningful to those involved and the resources available for a study. Included in the choice of methods should be a plan for public information about and discussion both of methods and the results of their application in planning.

REFERENCES

Carbonell, Armando J., Robert D. Yaro, et al., Bos-Nagymaros Barrage Study, Interim Report: Program Options and Impacts, Northampton, Mass., Ecologia, May, 1989.

Smardon, R. C., J. F. Palmer, and J. G. Felleman (eds.), Foundations for Visual Project Analysis, John Wiley and Sons, New York, 1986.

Zube, Ervin H., Robert O. Brush, and Julius Gy. Fabos, Landscape Assessment: Values, Perceptions, and Resources, Stroudsburg, Penn., Halsted Press, 1975.

Zube, Ervin H., James L. Sell, and Jonathan G. Taylor, "Landscape Perception: Research, Application and Theory," Landscape Planning 9, 1982, 1-33.

Chapter 5: The Demand Model

This chapter describes the demand model used in plan formulation for the NAR study. The demand model was used in conjunction with the supply model described in Chapter 6 to study the sources and costs of flow required to meet demands for water based on alternative sets of assumptions. The sections of the chapter include: an introduction; a description of model subroutines; a description of the mathematical forecasting relationships in the model; and perspectives on the model. The user's manual and detailed description of the model are in SP 29-219, and the data inputs to the model are described in detail in T-222 to T-270. The appendix to this chapter describes the derivation of the regional input-output model which is part of the demand model.

5.1 INTRODUCTION

The demand model is a group of subroutines based on forecasting relationships designed to yield estimates of demands for water based on specific assumptions. The mathematical components of the model are an input-output table for the NAR region, a regression estimator for publicly supplied municipal and industrial water, and a group of arithmetic operations associated with these components. The principal inputs to the model are projections of regional product, population, personal income, and water withdrawal coefficients, with projections of the geographic distribution within the region of these variables. The model acts on the inputs to produce estimates by benchmark years of water flow demands by economic sector, by type of water quality, and by subbasin, basin, state or area.

These flow demands are inputs to the supply model described in Chapter 6. The demands are net demands, in the sense that within the regional structure used for the model, gross water requirements in each subregion are reduced to account for reuse resulting from the distribution along a stream of the water users in the region. The regionalization is that shown in Figure 2-1 and described in NAR T-207 to T-220. (It should be noted that the term "demand model" can be replaced by "requirements model;" the demand model is not a demand model in the sense that price is an explicit variable in the estimating procedure.)

The model can be described in terms of four sets of information: the subroutines that were coded to permit the appropriate manipulation of input data; the mathematical forecasting operations incorporated within the subroutines; perspectives on the model; and the input data used. The first three sets of information are presented in this chapter; the fourth is available in T-222 to T-270. The general operation of the model is shown in Figure 5-1.

FIGURE 5-1: GENERAL SEQUENCE OF OPERATIONS OF THE DEMAND MODEL (Source: SP-39)

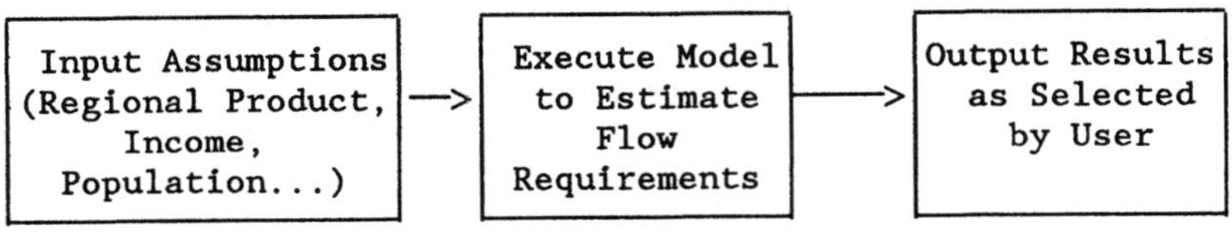

5.2 THE MODEL SUBROUTINES

The model has six subroutines called MUNWAS, INDACT, GROSSW, RURWAS, CONST, and NETW (for municipal water supply, industrial activity, gross water, rural water supply, constraint, and net water, respectively). The purpose of each subroutine is described in this section, and the various mathematical forecasting elements of the subroutines are described in the next section.

In subroutine MUNWAS, a regression estimator is used to generate estimates of future publicly supplied municipal and industrial water demands on the basis of projections for each subbasin of the population served and the per capita income.

In subroutine INDACT, dollar estimates of the gross output by economic sector by subbasin are generated on the basis of a regional input-output table, projections of regional final demand, and assumptions as to the distribution of sectoral gross output among the subbasins.

In subroutine GROSSW, estimates of future total gross water use by economic sector and by subbasin are derived. The dollar gross output estimates are multiplied by water use coefficients for different water qualities, and the totals are adjusted to remove the overlap between the regression estimates for publicly supplied water and the estimates of use based on the input-output table.

In subroutine RURWAS, future gross rural domestic water use is estimated on the basis of projections of the rural population and a per capita rural domestic use coefficient.

Subroutine CONST was written to permit the user to substitute exogenously derived water requirement estimates for any economic sector and subbasin for the estimates derived from the input-output procedure. This feature was added to permit a rapid comparison of the results of the estimates from the model with alternative estimates, providing for the incorporation of alternative estimates when these appear superior to the model estimates. Although the subroutine can be coded for any sector, for NAR planning it is coded in the model only for the agriculture and power sectors, the two sectors in which the exogenous estimates seemed to be potentially more accurate than the input-output estimates.

In subroutine NETW, the gross withdrawal demands for the various sectors are grouped into nine sectors which are useful in considering water demand and supply conditions (Figure 5-2). These aggregated

FIGURE 5-2: NET WATER DEMAND EXAMPLE (Source: Schaake & Major, 1972)

NET DEMAND (MGD) (NEFM(41,50,4))
NORTH ATLANTIC WATER RESOURCES STUDY
RUN = 20202RG1

RIVER BASIN MERRIMCK

WATER USE CATEGORY SUBBASIN MERRIMCK	1	2	3	4	5	6	7	8	9
FRESHWATER	65	58	0	3	17	31	483	140	34
WASTE WATER	0	4	0	0	0	0	0	4	0
BRACKISH WATER	0	0	0	0	0	0	0	0	0
CONSUMPTIVE	10	14	0	2	17	31	29	41	33

WATER USE CATEGORY SUBBASIN NSHMERMC	1	2	3	4	5	6	7	8	9
FRESHWATER	104	45	0	2	44	61	484	194	63
WASTE WATER	0	5	0	0	0	0	0	5	0
BRACKISH WATER	0	49	0	0	0	0	1564	49	0
CONSUMPTIVE	27	35	0	2	44	61	6	106	63

WATER USE CATEGORY	EXPLANATION
1	MUNICIPAL AND INDUSTRIAL (PUBLICLY SUPPLIED)
2	COMMERCIAL AND INDUSTRIAL (SELF SUPPLIED)
3	AGRICULTURAL SUPPLY (UNASSIGNED)
4	AGRICULTURAL SUPPLY (GROUNDWATER)
5	IRRIGATION (UNASSIGNED)
6	IRRIGATION (GROUNDWATER)
7	POWER COOLING
8	TOTAL UNASSIGNED WITHOUT COOLING
9	TOTAL ASSIGNED GROUNDWATER

sectoral gross demands are reduced to net demands for each subbasin to account for the fact that gross water demands occur at different points in a subbasin and that maximum flow required at a point is therefore less than the total of withdrawals throughout the subbasin.

5.3 MATHEMATICAL FORECASTING RELATIONSHIPS IN THE DEMAND MODEL

The mathematical forecasting relationships in the model are a series of simple algebraic equations and matrix manipulations.

Publicly supplied municipal and industrial water demand. Publicly supplied municipal and industrial demand for each subbasin is computed according to a regression equation derived from a time series analysis of information for Connecticut on the use of water, the population served, and per capita income. The regression estimator in the model was chosen over several others that gave statistical results approximately as good because it appeared to perform well in terms of the insights incorporated in handcrafted estimates for various NAR subbasins (R-6, R-7). The equation used in the model is:

$$Q = K(P)^{.825} x(Y)^{.308}$$

where: Q = public water requirement estimate in mgd;
P = population served in 1000's;
Y = per capita income in dollars; and
K = a constant that varies for each Area and is determined by conditions in the base year (R-7).

Industrial and other economic sector demands. Industrial and other sectoral water demands are estimated on the basis of an input-output table for the NAR and water withdrawal coefficients that relate economic activity to the water that is required to support that activity. The table used for the study is a regionalized version of a 1966 updating of the 1958 national table (Time, Inc., undated). The regionalization was carried out according to the methods of Lofting and Davis (1968) and Lofting and Davis (personal communication, 1968, 1969), and the matrix was reduced to 39 sectors, which corresponded to the availability of water use estimates. The water withdrawal coefficients are derived primarily from U.S. government publications, in particular the survey of water use in manufacturing that accompanied the 1963 Census of Manufactures (U.S. Bureau of the Census, 1966). The sources of water use coefficients are described in T-246 to T-270. An excellent introduction to input-output methods is Miernyk (1965); a comprehensive overview is given in Leontief (1986).

The model can be adapted to types of projections other than input-output projections; however input-output methods seem to be relatively good methods for a large region such as the NAR. The input-output framework gives consistency among the sectors for which projections are made, although the process of regionalizing the table, which involves making proportionality assumptions about national and regional relationships, tends to make this consistency more formal than empirical. The input-output table as used in the NAR as a projection device implies a constant pattern of interindustry transaction flows over time.

Sectoral gross outputs are projected for the NAR by multiplying the $(I - A)^{-1}$ matrix that is derived from the input-output transactions matrix (the A matrix) by a projection of regional final demand. This projection is distributed according to historical patterns or other desired patterns across the sectors of the input-output model. When regional final demand is multiplied by the $(I - A)^{-1}$ matrix, the total direct and indirect demands for the products of every sector are calculated. These outputs, called sectoral gross outputs, are distributed in the model to the 50 subbasins by a matrix of coefficients that gives the proportion of each sector that is assumed to be located in each of the subbasins. These proportions can be estimated by an analysis of information on employment or value added for each sector, or they may be projected on some other basis appropriate to the problem that is under consideration. The manipulations up to this point are performed in subroutine INDACT. In subroutine GROSSW, the estimates of the subbasin output for each sector are multiplied by coefficients for withdrawals of freshwater, brackish water, and waste water, and for consumption out of freshwater withdrawals. In addition, a matrix in this subroutine embodies estimates of the overlap between the input-output projections of water demand and the regression estimates; this overlap is removed in the subroutine to yield the final gross water withdrawal demands.

Estimates of rural domestic use. These estimates are derived by multiplying the estimated rural population in each subbasin by an estimated per capita water use coefficient. The rural population is computed as the difference between the total subbasin population and the municipally supplied population that is used as an input to the regression estimator for publicly supplied water. Rural domestic use is of interest as a separate use in the supply model because it is assumed in that model that this use is supplied by assigned groundwater rather than by withdrawals from surface sources.

Estimates of net water demand. The model up to the point of implementing subroutine NETW calculates the total withdrawal requirements for the NAR for all economic sectors. This information is of great interest, but for planning, the minimum total flow needed in each subbasin to supply withdrawal requirements is the most useful information. This figure is generally less than the sum of gross water withdrawal requirements for the several industries in a subbasin because the combined requirements can be met by the repeated use of the same water. To generate the flow demand that is required by the supply model, the 41 sectoral demands (39 input-output sectors and rural domestic and domestic) from the demand model are aggregated in the NETW subroutine to the nine categories shown in Fig. 5-2. These categories were chosen because the planners believed that estimates of the number of times each unit of water would be withdrawn in a given subbasin could be generated more easily in terms of these categories than in terms of the 41 categories for which the gross estimates are generated. A typical output from the demand model for the net water needs for a basin, inclusive of the nine categories, is given in Figure 5-2.

5.4 PERSPECTIVES ON THE DEMAND MODEL

5.4.1 COMPUTATIONAL ASPECTS. The demand forecasting system was designed within the context of computer technology at the time of planning so that the detailed operation of the model, the input information to the model, the output of information from the model, and the storage and retrieval of information from the data tapes are controlled by the user. The user controls the sequence of system operations by commands that are input to the system. Certain commands require additional data; a command and any required data for that command are called a command information set. The system will accept any sequence of command information sets that the user provides; therefore, the capabilities of the system include not only the individual commands that the system can execute but also more complex operations that may require a sequence of basic commands. This mode of operation is not typical of conventional Fortran program operation, and so the general structure of the system is not typical of other programs. For example, two users could solve identical problems with different sequences of commands. The advantage of this is that one or another sequence might be more appropriate for one or the other user in terms of his or her conceptualization of a problem or the organization of the required input information. This approach to the model, together with the output routines discussed in the next paragraph, illustrate the extent to which the NAR modeling effort prefigured the wide availability of user-friendly software. (For a review of the impact of changing computer technology on water resources modeling, see Loucks and Fedra, 1987.)

5.4.2 PLANNING ASPECTS. The model was designed to be used in a system of planning in which iteration between technicians and planners was stressed and carefully planned (Schwarz and Major, 1971). Hence the system was designed not only to perform appropriate computations but to respond to alternative needs for information in decision making. The user has the choice of what input information and output information is to be printed out for any run of the model. In addition, the model has the capacity to compute the algebraic differences between all inputs and outputs for any two runs. Computations are made according to a format in which all input and output elements that are unchanged between two runs print out as zeros. Therefore the output of this "compare" routine is ideally suited for planners who wish to scan outputs rapidly to detect and to comment on differences between alternative runs that emphasize, for example, different objectives or different technological assumptions. The printout information of the model (except for the listing of the Fortran statements) is formatted to government-sized paper, which in the United States is slightly smaller than commercial paper. Thus outputs can be reproduced directly for reporting, thereby eliminating the cost of typing and proofreading. Where appropriate, text for the output pages is provided.

It should be noted that the demand model information system was designed to be suitable for use in other planning situations. The system as used for the NAR could be duplicated with appropriate data and assumptions for other large-scale plans or for smaller regions where regional input-output tables are available. In addition, the system can be used as a flexible information manipulation system in which other types of projection are used. In such instances, alternative projections can enter the model at the appropriate point and earlier steps in the model as they apply to the NAR can be skipped. Further, the model can be used to compute the total pollution loadings for a given region by substituting the loading coefficients of different types of pollution for the withdrawal coefficients presently used in the model.

5.4.3 ALTERNATIVE METHODS. The principal elements in the demand model, the regionalized input-output table and the generalized regression estimator for publicly supplied municipal and industrial water, provided an appropriate basis for forecasting at the framework level in the NAR. In other regional studies, different approaches may be desirable. (A study of the application of statistical demand estimation to urban water management is Jones et al., 1984.)

For example, where detailed forecasts of residential water use are required (as was not the case in the NAR) the regression equations found in Howe and Linaweaver (1967) and Howe (1982) can be considered. These equations were developed from a detailed data set collected by Johns Hopkins University's "Residential Water Use Research Project" (Linaweaver et al., 1966). Metered and sewered areas in the eastern and western United States, metered areas with septic tanks, apartments and flat-rate (unmetered) areas were examined separately to produce the Howe-Linaweaver equations. These were adapted for, and are conveniently available in, the IWR-MAIN forecasting model sponsored by the U.S. Army Corps of Engineers; the user's manual is available as Davis et al. (1988). The IWR-MAIN system is designed to be used with an IBM PC/XT/AT or compatible microcomputer, and a system diskette is available from the Corps of Engineers group that maintains the model and publishes the manual. The IWR-MAIN model shares with the NAR the attempt to incorporate the mathematical elements of forecasting into a computational framework that is user-friendly.

When an input-output table is not available for a region, an alternative method often employed for nonresidential water use forecasting is to project employment by sector and then to multiply the employment forecasts by water use per employee coefficients. This is the method used in the IWR-MAIN model; the manual includes lists of coefficients that can be considered for various sectors.

Regional water use for all sectors, nonresidential as well as residential, can be forecast using econometric methods. One example of the use of these techniques is in the Seattle, Washington area. The Seattle Water Department, a regional supplier of 1.1 million people, uses regression equations for both short and long term forecasting. An insightful analysis of the Seattle methods in the context both of the development of required data and the purposes of water use forecasting is provided by Jones et al. (1984) and DeKay (1985).

REFERENCES

Davis, William Y., et al., IWR-MAIN Water Use Forecasting System, Version 5.1: User's Manual and System Description, U.S. Army Corps of Engineers, Water Resources Support Center, Institute for Water Resources, IWR Report 88-R-6, Fort Belvoir, Va, 1988

DeKay, C. Frederick, "The Evolution of Water Demand Forecasting," Journal of the American Water Works Association, October, 1985, 54-61.

Howe, Charles W., and F.P. Linaweaver Jr., "The Impact of Price on Residential Water Demand and its Relation to System Design and Price Structure," Water Resources Research 3:1, 1967, 13-32.

Howe, Charles W., "The Impact of Price on Residential Water Demand: Some New Insights," Water Resources Research 18:4, 1982, 713-716.

Jones, C. Vaughan, et al., Municipal Water Demand: Statistical and Management Issues, Westview Press, Studies in Water Policy and Management No. 4, Boulder,Col., 1984.

Leontief, Wassily, Input-Output Economics, New York: Oxford University Press, 2nd ed., 1986.

Linaweaver, F.P. Jr., J.C. Beebe, and F.A. Skrivan, Residential Water Use Research Project, Report IV, Phase 2, Data Report, Department of Environmental Engineering Sciences, John Hopkins University, Baltimore, Md., June, 1966.

Lofting, E.M., and H.C. Davis, "The Interindustry Water Matrix: Applications on a Multiregional Basis," Water Resources Research 4:4, 1968, 689-695.

Lofting, E.M., and H.C. Davis, Personal Communications to NAR Staff, 1968, 1969.

Loucks, Daniel P., and Kurt Fedra, "Impact of Changing Computer Technology on Hydrologic and Water Resources Modeling," Reviews of Geophysics 25:2, March, 1987, 107-112.

Miernyk, W.H., The Elements of Input-Output Analysis, Random House, New York, 1965.

Schaake, John C. Jr., and David C. Major, "Model for Estimating Regional Water Needs," Water Resources Research 8:3, June, 1972, 755-759.

Schwarz, H.E. and David C. Major, "An Experience in Planning: The Systems Approach," Water Spectrum, 3:3, Fall, 1971, 29-34.

Time, Inc., Fortune's 1966 Input/Output Coefficients, New York, undated.

U.S. Bureau of the Census, Census of Manufactures, 1963. Subject statistics. Water use in manufacturing. Subj. Rep. MC63(1)-10, U.S Government Printing Office, Washington, D.C., 1966.

Appendix to Chapter 5: Derivation of the Regional Input-Output Model

The model used in the NAR study (T-236 to T-241) focuses on the structure of a particular region and its relationships with all other areas (referred to as the rest of the world). The regional direct coefficients are derived by adjusting national coefficients to reflect the spatial quality of regional production. In order to apply the procedure, two key assumptions must be made: (1) regional technology is identical to national technology; and (2) the region only imports when it has exhausted its own resources.

The first assumption is usually made in order to eliminate the immense task of developing actual technological data for a region. It will be close to reality for a large diversified region; that is, the national coefficients can be thought of as a weighted average of regional coefficients and thus a large region would tend to influence national coefficients to a great extent. Adjustments for regional industrial mix are possible and simple computationally, but difficult to do without access to census data banks.

The second assumption enables the model to provide a regional direct coefficient matrix suitable for projections. It implies that the level of exports and imports for a region can be derived given the levels of gross output, intermediate demand, and final demand. With the level of imports by industry and some scheme to allocate the total amount of imports among the purchasing industries, the regional direct coefficient matrix is easily calculated. To clarify the exposition, a numerical example will be introduced. The example begins with a national flow table, values for national gross output, regional gross output, and regional final demand. Regional gross output and national final demand can be estimated from national gross output and national final demand by determination of the proportion of national value added produced in the region and the proportion of national personal income earned in the region. That is, if:

X_i = national gross output industry i
XR_i = regional gross output industry i
VA_i = national value added industry i
VAR_i = regional value added industry i

then,

$$XR_i = \frac{VAR_i}{VA_i} (X_i)$$

This assumes that gross output is spatially distributed as value added, and if:

Y_i = national final demand industry i
YR_i = regional final demand industry i
PI_i = national personal income industry i
PIR_i = regional personal income industry i

then

$$YR_i = \frac{PIR_i}{PI_i}(Y_i) \quad ,$$

with a similar assumption about final demand (household demand in this case) and personal income. Other measures such as sales and government employment can also be utilized where appropriate. The national figures are available from national tables published by the U.S. Department of Commerce, Office of Business Economics. The regional values are derived as follows:

Let: YR = regional final demand
XR = regional gross output vector
X = national gross output vector
A = national flow matrix.

Also let: XR = (10 40 20 10)
X = (100 200 200 400)

$$A = \begin{bmatrix} 20 & 0 & 0 & 40 \\ 0 & 40 & 80 & 40 \\ 0 & 60 & 0 & 0 \\ 20 & 0 & 0 & 200 \end{bmatrix}$$

YR (17 3 8 9)

The national coefficient matrix is derived first by dividing each column by the gross output of the industry represented in the column. This is done by postmultiplying the national flow matrix by the inverse of a diagonal matrix with national gross output along the diagonal.

Let: $\bar{A}$ = national direct coefficient matrix
X^d = diagonal matrix formed from national gross output vector.

$\bar{A} = A\,(X^d)^{-1}$, then

$$\bar{A} = \begin{bmatrix} .20 & 0 & 0 & .10 \\ 0 & .20 & .40 & .10 \\ 0 & .30 & 0 & 0 \\ .20 & 0 & 0 & .50 \end{bmatrix}$$

The .30 in row 3 column 2 will therefore mean that to produce one unit of industry 2's output requires .3 units of industry 3's output.

Next a regional flow table is derived that represents the flow that would be observed in the region, assuming regional technology and trade structure to be identical to the national counterparts. That is, this table or matrix assumes the region merely to be a scaled down version of the nation. This first approximation to the desired table is constructed by postmultiplying the national direct coefficient matrix by a diagonal matrix formed from the regional gross output vector.

Let: RF = regional flow matrix (non-adjusted)
XR^d = diagonal matrix formed from regional gross output vector

$RF = \bar{A}.XR^d$, then

$$RF = \begin{bmatrix} 2 & 0 & 0 & 1 \\ 0 & 8 & 8 & 1 \\ 0 & 12 & 0 & 0 \\ 2 & 0 & 0 & 5 \end{bmatrix}$$

This matrix yields the intermediate flow necessary to produce the given regional gross output using national technology; it gives no indication as to the source of these flows, that is, whether they are internal to the region or imported from other regions.

The rows of the non-adjusted regional flow matrix are summed to obtain intermediate output necessary to produce the given gross output. This intermediate output plus the level of final demand yields the level of output necessary to support regional activity. If this is compared to the level of regional gross output, it can be determined whether the region produces enough itself to support the level of its activity. Using the assumption that the region only imports when it has exhausted its own resources, we compute a vector called net trade balance by subtracting intermediate plus final demand from regional gross output. Positive levels of net trade balance indicate that an industry that produces more than is necessary to support its level of intermediate and final demand will export. Negative levels for net trade balance represent industries that must import. The net trade balance vector is formed and then broken into two vectors, commodity imports and commodity exports. Negative net trade balances are imports and positive balances exports.

Let: NETB = net trade balance vector
CE = commodity export vector
CI = commodity import vector, then

$$NETB = (I-\bar{A})\ XR-YR$$

$$NETB = \begin{bmatrix} 10 & -3 & -17 \\ 40 & -17 & -3 \\ 20 & -12 & -8 \\ 10 & -7 & -9 \end{bmatrix} = \begin{bmatrix} -10 \\ 20 \\ 0 \\ -6 \end{bmatrix}$$

$$CE = (\;0 \quad 20 \quad 0 \quad 0)$$
$$CI = (-10 \quad 0 \quad 0 \quad -6)$$

The level of imports is thus derived.

In order to derive the adjusted regional direct coefficient matrix it is necessary to make an assumption about the allocation of imports among industries. The usual and also most intuitively pleasing assumption is that imports are distributed in the same proportion as flows of gross output. For example, if the flow from industry 1 to industry 4 is one tenth the gross output of industry 1, it can be assumed that industry 1 accounts for one tenth of the imports of industry 1 of industry 4. This in effect says that the imports are identical to the product produced by the industry, and each purchaser buys imports in relation to its importance as a purchaser. This brings to light the problem that derives from the fact that the region is treated as if it were a single dimensionless point. For a single point there would be no reason to believe that certain purchasers would use greater portions of imports than their output proportion indicates. In reality, industries located at the periphery of the region, other things being equal, could be expected to purchase larger portions of imports than anticipated. Also exports would be possible even in an industry that does not produce enough domestically to supply intermediate and final demand. This would be conceivable as industries at the periphery of the region may supply their whole requirements from outside sources, thus releasing internally produced commodities for export from other parts of the region. This problem, due to lack of spatial information about the region itself, is not peculiar to this method and applies to any model that must handle a dimensional entity as a nondimensional point.

Returning to the example, the assumption is accepted that imports are distributed according to flow proportions of gross output, and a matrix is derived indicating these proportions.

Let: $\bar{R}$ = row percentages
X_{ij} = flow from i to j
Y_i = final demand industry i, then

$$\bar{R} = X_{ij} / \sum_j (X_{ij} + Y_i)$$

$$\bar{R} = \begin{bmatrix} .10 & 0 & 0 & .05 \\ 0 & .40 & .40 & .05 \\ 0 & .60 & 0 & 0 \\ .1250 & 0 & 0 & .3125 \end{bmatrix}$$

Multiplying this matrix by the absolute level of imports yields an import matrix showing absolute levels of imports for each cell of the regional flow matrix.

Let: M = import matrix

CI^d = diagonal matrix with CI vector along diagonal,

then $M = CI^d.\bar{R}$

$$M = \begin{bmatrix} 1 & 0 & 0 & .50 \\ 0 & 0 & 0 & 0 \\ 0 & 0 & 0 & 0 \\ .750 & 0 & 0 & 1.875 \end{bmatrix}$$

Subtracting this matrix from the regional flow matrix yields an adjusted regional flow matrix representing the internally produced flows for the region. If we let:

$\overline{RF}$= adjusted regional flow matrix

RF= RF - M; then

$$\overline{RF} = \begin{bmatrix} 1 & 0 & 0 & .50 \\ 0 & 8 & 8 & 1 \\ 0 & 12 & 0 & 0 \\ 1.250 & 0 & 0 & 3.125 \end{bmatrix}$$

This means that, for instance, of the 2 units of flow from industry 1 to industry 2, one of the units is produced by industry 1 in the region and one unit by industry 1 outside the region.

Deriving adjusted regional direct coefficients by dividing each column of RF by the appropriate level of output yields a coefficient matrix that can be used to provide the level of output produced in the region necessary to support any level of final demand.

Let: $\overline{AR}$ = adjusted regional direct coefficient matrix

X_{ij} = adjusted flow from i to j

$$\overline{AR} = \begin{bmatrix} .1000 & 0 & 0 & .0500 \\ 0 & .2000 & .4000 & .1000 \\ 0 & .3000 & 0 & 0 \\ .1250 & 0 & 0 & .3125 \end{bmatrix}$$

Chapter 6: The Supply Model

This chapter describes the supply model used in plan formulation for the NAR. As described earlier the supply model was used to estimate sources and costs of flow to meet the demands specified by the NAR demand model. The sections of the chapter include: an introduction; model formulation; design variables and constraints; the objective function; and perspectives on the model. The model is described in detail in SP 220-289. Information on mathematical programming in general is available in many texts, among them Hadley (1962), Wagner (1975), and Bradley et al. (1977). Applications of mathematical programming to water resources are discussed in Maass et al. (1962), Hall and Dracup (1970), Buras (1972), and Loucks et al. (1981); a more general treatment of applications is in Cohon (1978). Linear and mixed-integer programming algorithms are now widely available for microcomputers; examples are LP83 and MIP83 (Sunset Software, 1987).

6.1 INTRODUCTION

The supply model is a mathematical programming model for plan formulation designed to permit evaluation of the sources and cost of the supplies of water required to meet various specified subarea water requirements. The basic inputs to the model are withdrawal, instream, and consumption requirements for the 50 NAR subareas and combinations thereof; data on existing and potential intrabasin and interbasin transfers; and data on the costs of development. Provision is made for inserting and varying parametrically constraints on the source, degree, and type of development to be used in the model.

The classification of the water types in the model is compatible with the classification scheme of the NAR demand model. The model considers two qualities of water, freshwater of class D (water usable with complete treatment) or better and waste water (water used directly from a waste water source, not acceptable for freshwater use).

In each subbasin some groundwater demands are assigned. Thus a certain portion of the groundwater storage developed in the subbasin must be used for specified activities, such as irrigation and rural water supply. Figure 6-1 presents a schematic representation of the NAR region with natural drainage connections and potential water transfers.

6.2 MODEL FORMULATION

In the model the following simplifying assumptions were made:

FIGURE 6-1: NAR REGIONAL SCHEMATIC OF EXISTING AND POTENTIAL WATER TRANSFER NETWORK (Source: SP-224)

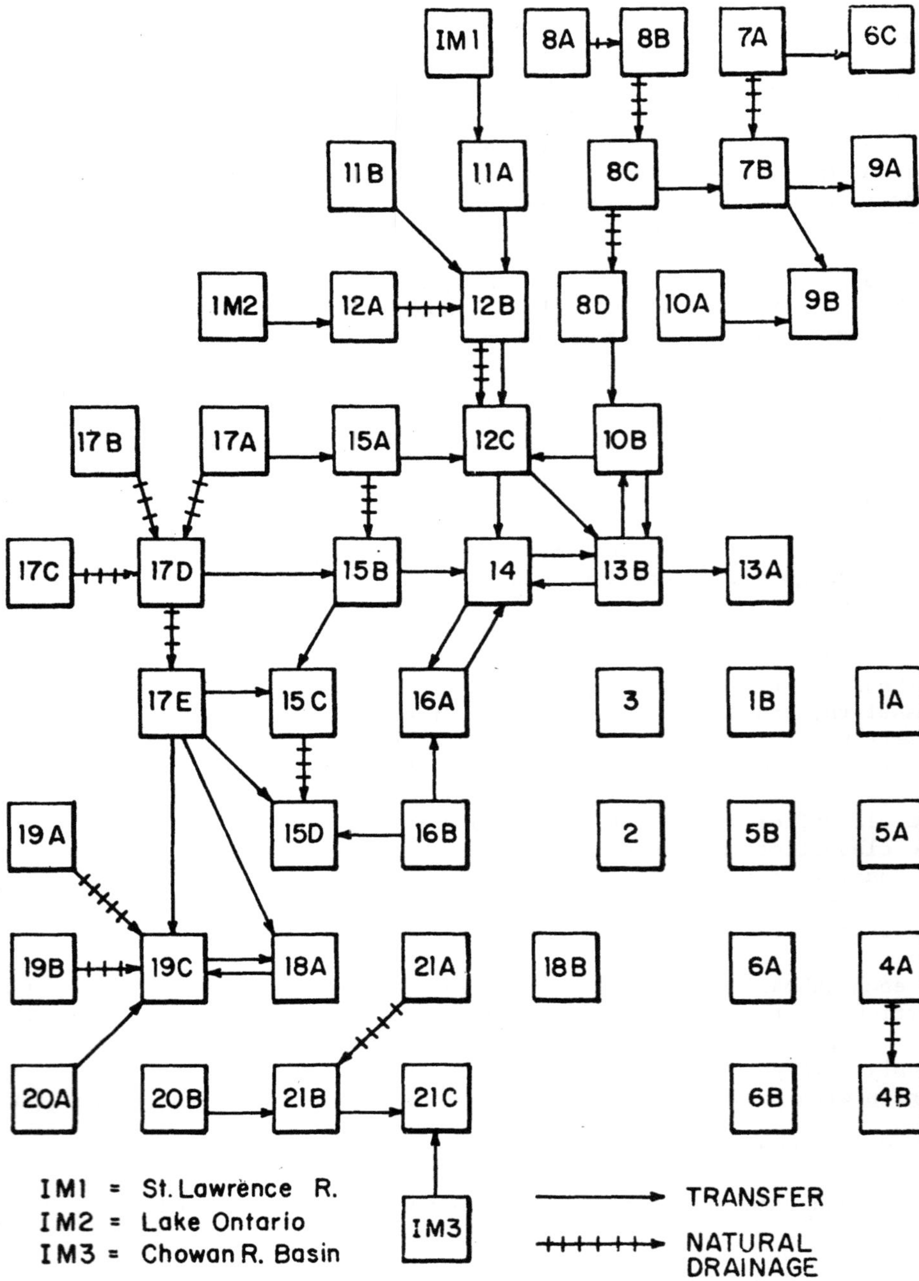

1. All the water supply and water use activities in a subasin are considered to be located at the same point. Therefore the NAR becomes a series of 50 nodes that are linked by natural drainage and human-made interbasin transfers.
2. All yields from storages are expressed in annual terms converted into million gallons per day (mgd). The formulation is a static annual deterministic one that is based on critical period analyses and selected risk levels as described in Chapter 7.
3. Return flows from freshwater can be used to satisfy demands for wastewater.
4. Interbasin groundwater movements have been ignored.

The first assumption allows generalized continuity relationships to be written for each subbasin. A typical subbasin is represented schematically in Figure 6-2. The second assumption is used to aggregate the mainstream storage that has a safe yield of S_i. Upstream storage, groundwater storage, and a flow generated within the subbasin are similarly considered with the respective yield variables U_i, G_i, and Z_i. Groundwater yield is divided into assigned A_i and unassigned N_i. Figure 6-2 shows the water that is available for meeting the demands in each subbasin entering a hopper. These supplies include, in addition to unassigned groundwater, the yield of stream withdrawals R_i, desalinized saline or brackish water B_i, and imports from subbasin j, I_{ij}. The demand for assigned groundwater D_{Ai}, is met directly by the assigned groundwater yield. The other supplies are then assumed to be available to meet the demands for freshwater D_{Fi}, or to be used for export to j, E_{ij}.

The third assumption allows the wastewater demand D_{Wi} to be met from the return flow from freshwater. The fourth assumption relates the subbasins to each other by the continuity of the natural surface drainage system variables of subbasin inflow Y_i, and outflow X_i , as well as by imports and exports.

The supply model accepts as parameters from the demand model the demands for freshwater and wastewater and the consumptive losses of freshwater L_{Fi}, waste water L_{Wi}, and assigned groundwater L_{Ai}, in each subbasin.

6.3 DESIGN VARIABLES AND CONSTRAINTS

The model is explained algebraically in this section. The design variables of the model are:

U_i	yield from upstream storage in i
S_i	yield from mainstream storage in i
G_i	yield from groundwater storage developed in i
A_i	assigned yield from groundwater storage in i

FIGURE 6-2: TYPICAL SUBBASIN i (Source: SP-225)

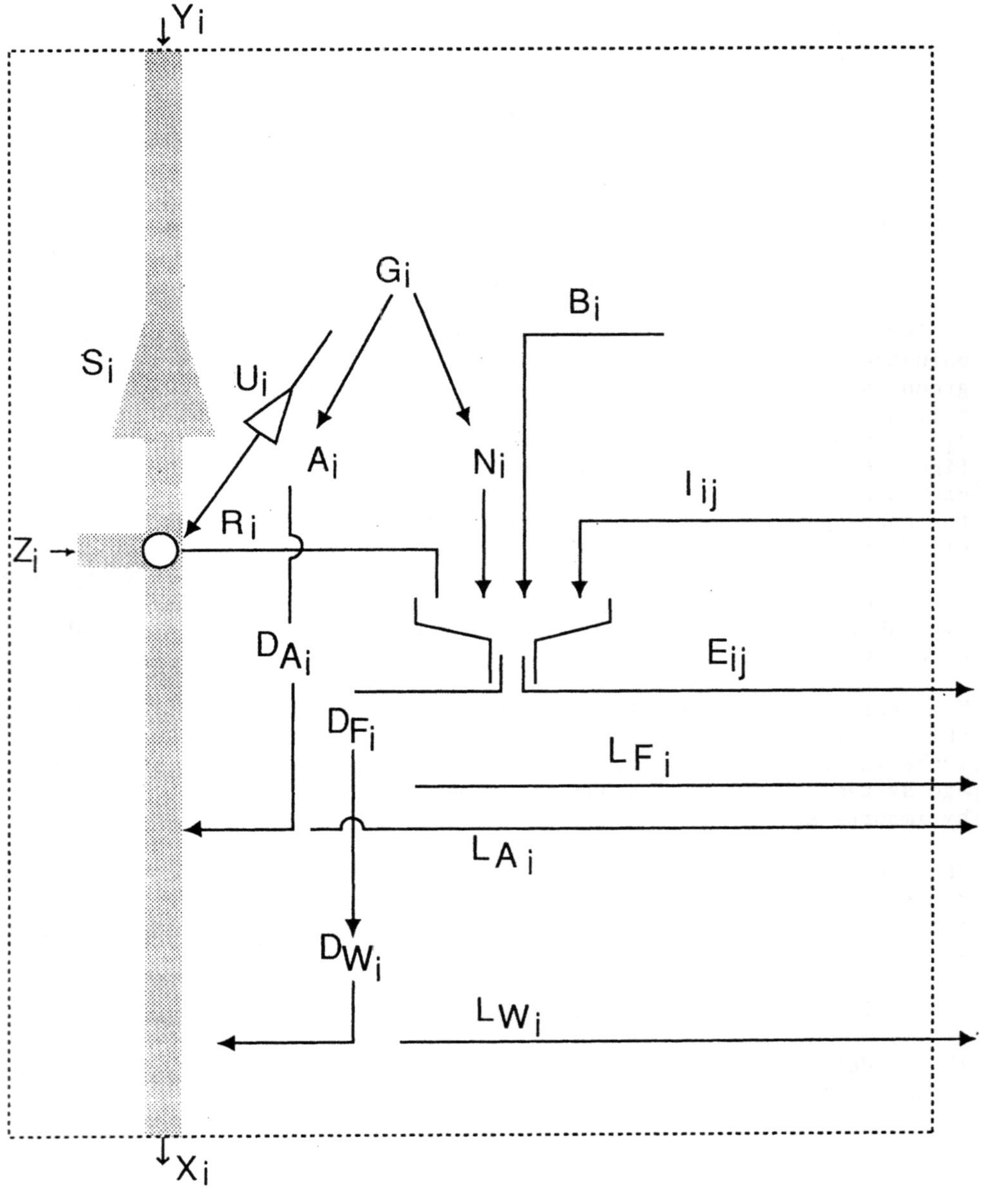

The model constraints include:

1. Upper limits on yields from groundwater, upstream and main-stream storage, and a limit on total subbasin yield.
2. Low flow constraints at the downstream end of the sub basin (P_i) and low flow constraints within the subbasin (Q_i).
3. Continuity or mass balance at each subbasin.
4. Freshwater, wastewater, and assigned groundwater demand constraints.
5. Equality relation between import-export coupled pairs.
6. Upper limits on desalinated water and on capacities for export and import.
7. Lower limits on existing transports.
8. A relationship between groundwater development and available surface flows that are generated in a subbasin, with the surface flow generated by having an upper limit.
9. Continuity on interbasin flow for natural and human-made transports.

These constraint sets, with the variables and constraints as defined earlier, are written below (deLucia and Rogers, 1972).

Constraint 1

$$U_i \leq \bar{U}_i \qquad \forall\ i$$
$$S_i \leq \bar{S}_i \qquad \forall\ i$$
$$N_i + A_i = G_i \leq \bar{G}_i \qquad \forall\ i$$

Constraint 2

$$X_i \geq P_i \qquad i \in I_k$$
$$Y_i + Z_i + S_i + U_i - R_i \geq Q_i \qquad \forall\ i$$

Constraint 3

$$Y_i + Z_i + G_i + U_i + S_i + B_i + \sum_{j \in J_i'} I_{ij} = X_i + \sum_{j \in J_i} E_{ij} + L_{F_i} + L_{W_i} + L_{A_i} \qquad \forall\ i$$

Constraint 4

$$\sum_{j \in J_i'} I_{ij} + R_i + N_i + B_i \geq D_{F_i} + \sum_{j \in J_i} E_{ij} \qquad \forall\ i$$
$$\sum_{j \in J_i'} I_{ij} + R_i + N_i + B_i \geq L_{F_i} - D_{W_i} + \sum_{j \in J_i} E_{ij} \qquad \forall\ i$$
$$A_i = D_{A_i}$$

Constraint 5

$$I_{ij} = E_{ji} \qquad \forall\ i, j \in \lambda$$

Constraint 6

$$B_i \leq \bar{B}_i$$
$$E_{ij} \leq \bar{E}_{ij} \qquad \forall\ i \text{ and } j \in J_i$$
$$I_{ij} \leq \bar{I}_{ij} \qquad \forall\ i \text{ and } j \in J_i'$$

Constraint 7

$$E_{ij} \geq E_{E_{ij}} \qquad \forall\ i \text{ and } j \in K_i$$
$$I_{ij} \geq I_{E_{ij}} \qquad \forall\ i \text{ and } j \in K_i'$$

Constraint 8

$$Z_i = \bar{Z}_i - \alpha_i N_i \qquad \forall\ i$$
$$Z_i \leq \bar{Z}_i$$

Constraint 9

$$\sum_{i \in M_i} X_i = Y_{i+1}$$

The constraint sets, with the exception of 2, reflect physical relationships or resource limitations. Constraint 2 is included to capture legal and other requirements on the downstream flow levels, and to attempt, within the limitations of treating subbasins as nodes, to reflect quality and recreational objectives by forcing flow "within the node" to be above a certain level. (These "within the node" constraints are very aggregate in nature and are subject to the usual limitations of aggregate measures.)

6.4 OBJECTIVE FUNCTION

The objective is to minimize the total annual costs of meeting the constraints. However, provision is made for inserting and varying parametrically constraints on sources, degrees, and types of development in the model, and in this way the influence of planning objectives other than cost minimization can be brought to bear on model results.

Costs in the objective function (with the "C" terms defined as annual costs per mgd of yield) are the sum of costs for imports, mainstem storage, upstream storage, assigned and non-assigned groundwater development and desalinization, or

$$\sum_{j \in J_i'} \sum_i C_{ij}(I_{ij}) + \sum_i C_{S_i}(S_i) + C_{U_i}(U_i) + C_{A_i}(A_i) + C_{N_i}(N_i) + C_{B_i}(B_i)$$

The objective function is a sum of linear and nonlinear cost functions; the nonlinear functions are often non-convex. Figure 6-3 presents two examples of the cost curves used, together with the piecewise linear approximations used to represent nonlinear curves in the model. The cost curves were developed by the U.S. Army Corps of Engineers and the other cooperating agencies in the NAR study. In this model an operating policy, interest rates, time horizons, and risk levels were assumed and the relevant cost yield curve was derived. Alternative cost functions can readily be incorporated into the model. The minimum cost efficiency objective function could be modified by including additional relevant costs or benefits associated with any of the model variables. One possibility is the treatment of demands as variables in constraint 4 rather than as input parameters. Benefit estimates could be attached to the demand variables and included in the objective function. If benefit estimates of the demands were not available, demands could still be allowed to vary within the bounds of the present demand estimates generated by the demand model.

The model yields a great deal of information with respect to the demand and other constraints. In addition to producing the optimum set of supplies and the costs thereof, the model solution produces information on the shadow prices or costs associated with binding constraints and bounds. (For a thorough treatment of the theory of the dual problem including extensive discussions of the simplex multiplier [dual activity] see Dantzig, 1963.) This information is contained in the dual

FIGURE 6-3: COST CURVE FOR MAIN STEM RESERVOIRS (Source: SP-230)

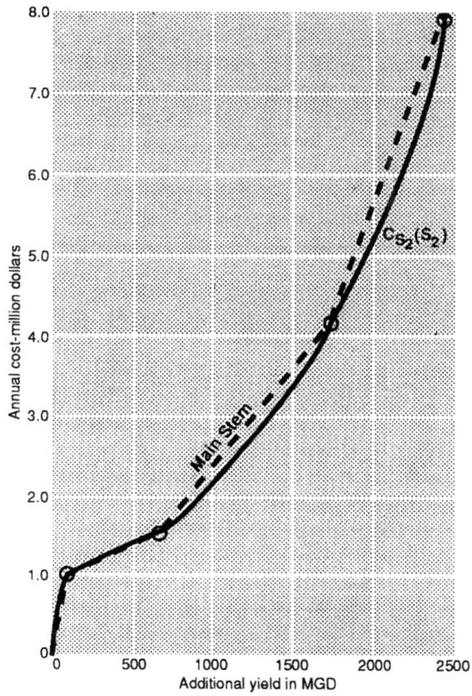

COST CURVE FOR WATER TRANSFERS (Source: SP-232)

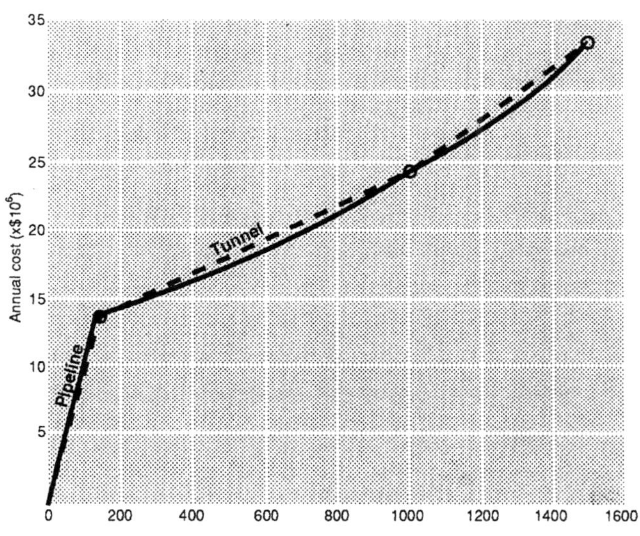

activity and reduced cost output values of IBM's Mathematical Programming System (IBM, 1972)

These values are the rate of change of the objective function's value per unit of change in the associated right hand side of the constraint. For example, the appropriate dual value indicates how much system costs (in annual terms) would decrease from a unit decrease in the (binding) freshwater demand in a subbasin. Thus, by comparing the dual values of binding demand constraints, the relative system costs of these demands can be examined.

6.5 PERSPECTIVES ON THE SUPPLY MODEL

6.5.1 COMPUTATIONAL CONSIDERATIONS. The NAR supply model is a mathematical programming model with a nonlinear objective function and all linear constraints. The model was programmed by using a separable programming algorithm. The IBM MPS/360 system that was used employed the 'delta' (Hadley, 1964) form of the separable algorithm for approximating the nonlinear functions by piecewise linear functions. (Nonconvex properties of the objective function and the need to use a number of program input/output and post-optimal options argued against the use of available network programs.) The piecewise linear approximations for some example cost curves are shown in Figure 6-3.

The resulting model, after the inclusion of separable approximating variables and equations, has a matrix of over 2000 variables and 1000 rows. While the NAR plan was being formulated, the model was run numerous times; demand sets corresponded to the various benchmark years and objectives, and model parameters reflected different low flow objectives.

Because some of the cost functions are non-convex the objective function may be non-convex in the feasible domain, and therefore an optimal solution generated may be a local optimum (Hadley, 1964, p. 8). A series of heuristics was used to deal with this and other difficulties. Multiple starts and a selected use of bounds were used to identify additional local optima. In the course of developing series of runs for different benchmark years judgement was exercised in selecting starting points for computer runs and constraints on certain activities to reflect information that was generated by earlier runs.

6.5.2 MODEL USE AND FORMULATION. In an attempt to facilitate the use of the model in plan formulation, a series of executive output routines was programmed that display the programming solution output in an immediately useful form. An example of this output format is shown in Table 6-1. Additional routines were developed to compare the results of different runs and to present them in similar format, to aggregate from subbasins to basins, and to transfer the development costs into implicit budgets within the planning periods. The executive output routines developed for the supply model, like those for the demand

TABLE 6-1 (Source: SP-246)

BASIN 6C PRES SAC SUB-BASIN PISCATQA RUN YEAR 2000

SOURCES OF FLOW AND COST TO MEET FRESH WATER DEMAND 69 MGD, WASTE WATER DEMAND 1 MGD, INSTREAM DEMAND 82 MGD, AND OUTFLOW MINIMUM 103 MGD.

SOURCE	FLOW (MGD)	ANNUAL COST ($MILLION 1970)
UPSTREAM STORAGE	34	0.20
MAINSTREAM STORAGE	0	0.00
GROUNDWATER ASSIGNED	12	0.32
GROUNDWATER UNASSIGNED	0	0.00
DESALTED WATER	0	0.00
EXISTING SUB-BASIN INTRAFLOW	105	0.00
SUBTOTAL	151	0.52
IMPORT FROM SUB-BASIN MERRIMACK	0	0.00
NET TRANSFER	0	
TOTAL FLOW AVAILABLE	151	
CONSUMPTIVE LOSS	18	
SUB-BASIN OUTFLOW	133	

model, were among the most important methods of communication developed for the NAR study.

The supply model was formulated with an objective function that minimized national income costs (other objectives being reflected in the constraints), and with all linear constraints. This formulation worked well, in general, given the data and methods used in the NAR study. Alternative formulations must be considered for every new study. For many studies, formulations with explicit net benefit maximization (either for a single objective or for multiobjectives) will yield improved results, as will the use of integer as well as linear constraints (mixed-integer programming). Examples of models with both features are described in Major and Lenton, 1979, chs. 5 and 8.

Because of the power of mathematical programming and the availability of suitable algorithms for microcomputers, it can be expected that water resources agencies will make increasing practical use of the technique in planning. The right formulation for a particular problem is not by any means always evident early in a study. Some of the highways and byways in the development of models for a study are described in Major and Lenton (1979, app. A). Rogers (1978) provides perspectives on the issue of simple versus complex models.

REFERENCES

Bradley, Stephen P., A. C. Hax, and T. L. Magnati, Applied Mathematical Programming, Addison-Wesley, Reading, Mass., 1977.

Buras, Nathan, Scientific Allocation of Water Resources, American General Services Publishing Company, New York, 1972.

Cohon, Jared, Multiobjective Programming and Planning, Academic Press, New York, 1979.

Dantzig, George, Linear Programming and Extensions, Princeton University Press, Princeton, N.J., 1963.

deLucia, Russell J. and Peter Rogers, "North Atlantic Regional Supply Model," Water Resources Research 8:3, June, 1972, 760-764.

Hadley, G., Linear Programming, Addison-Wesley, Reading, Mass., 1962.

Hadley, G., Non-Linear and Dynamic Programming, Addison-Wesley, Reading, Mass., 1964.

Hall, Warren A. and John A. Dracup, Water Resources Systems Analysis, McGraw-Hill, New York, 1964.

IBM, "Mathematical Programming System--Extended (MPSX): Program Description'" TNLSH 20-0968-1, IBM Corporation, White Plains, New York, 1972.

Loucks, D. P., J. R. Stedinger, and D. A. Haith, Water Resources Systems Planning and Analysis, Prentice-Hall, Englewood Cliffs, N.J., 1981.

Maass, Arthur et al., Design of Water-Resource Systems, Harvard University Press, Cambridge, Mass., 1962.

Major, David C., and Roberto L. Lenton, Applied Water Resource Systems Planning, Prentice-Hall, Englewood Cliffs, N.J., 1979.

Rogers, Peter, "On the Choice of the 'Appropriate Model' for Water Resources Planning and Management," Water Resources Research 14:6, December, 1978, 1003-1010

Sunset Software, "LP83/MIP83: A Professional Linear Programming System," San Marino, California, 1987.

Wagner, Harvey M., Principles of Operations Research, Prentice-Hall, Englewood Cliffs, N.J., 2nd ed.,1975.

Chapter 7: The Storage-Yield Model

This chapter describes the NAR storage-yield model, together with the shortage index associated with the model in terms of which failures are expressed. The storage-yield relationships used in the supply model are based upon the analyses conducted with the model described here. The model was designed to determine the amount of storage required and the risk of failure of that storage at various demand rates at streamflow locations where the flow is known, and to express the failures in terms of a shortage index. A detailed description of the model is in SP-1 through SP-28; hydrology in the NAR is discussed in App. C.

7.1 THE STORAGE-YIELD MODEL AND SHORTAGE INDEX

In the storage-yield model, monthly historical streamflows for groups of stations are extended to equal length and synthetic streamflows of 100 years length per series are generated (A standard reference on synthetic streamflows is Fiering and Jackson, 1971.) The U.S. Army Corps of Engineers Hydrologic Engineering Center Computer Program 23-C*-L267, "Monthly Streamflow Simulation," is used for these operations. These streamflows are then routed past a station at predetermined storage increments to determine the shortages obtained given pre-set yield requirements. Deficiencies are expressed in terms of the shortage index.

The input set consists of monthly historic flow adjusted to natural conditions, and ten separate series of monthly generated flows of 100 years, each in groups of up to ten gaging stations. Historic records were extended to the length of the station with the longest historic record in each group.

The program simulates the operation of a reservoir for each gaging station and for the local area between gaging stations by applying the balance equation:

$$S_{fi} = S_i + (I - O)$$

where:

S_{fi} = the storage at the end of month i
S_i = the storage at the end of the previous month
I_i = average inflow during month i
O_i = average outflow during month i

Local area streamflows for differential areas between gages are determined by subtraction of flows between the downstream gage and the successive upstream gage or gages.

If S_{fi} becomes negative in the above equation, a shortage exists and continues for the duration that 0 exceeds I. The magnitude of the shortages expressed as a shortage index is computed using the equation:

$$S.I. = \left[\sum_{1}^{N} \left[\frac{\sum_{i=1}^{12} (-S_{fi})}{\sum_{i=1}^{12} (O_i)} \right]^2 \right] \frac{100}{N}$$

where: N = number of years of record
i = month number

The shortage index is explained below. All negative and zero end-of-the-month storages are used to determine the number and duration of failure-months for each storage amount and draft rate during each period; an example is shown in Figure 7-1.

7.2 APPLICATION OF MODEL AND INDEX

The draft rates, selected as percentages of average flow, were held constant at 20, 40, 50, 60, 70, 80 and 90 % intervals while the storage was increased until one or no failures occurred. Provision was made for monthly variation in demand based on observed variability of demand in the North Atlantic Region, as shown here:

JAN	FEB	MAR	APR	MAY	JUN	JUL	AUG	SEP	OCT	NOV	DEC
95%	95%	95%	95%	100%	110%	110%	110%	105%	95%	95%	95%

All draft rates are initially routed at zero storage. The second routing sequence begins with an assumed storage dependent upon the size of the drainage area and draft rate. There are three cases. In Case I, where one inch of storage over the drainage area is equivalent to less than 100,000 acre-feet, the storage for the second routing sequence is set equal to one-half inch at the 20 % draft rate, one inch at 40 % and 50 %, two inches at 60 %, 70 % and 80 %, and three inches at 90 %. In Case II, where one inch of storage is between 100,000 and 200,000 acre-feet, the second routing sequence storage assumption is 50,000 acre-feet at the 20 % draft rate, 100,000 acre-feet at 40 % and 50 %, 200,000 acre-feet at 60 %, 70 % and 80 %, and 300,000 acre-feet at 90 %. In Case III, where one inch is greater than 200,000 acre-feet, the second sequence is 100,000 acre-feet at 20 %, 200,000 acre-feet at 40 % and 50 %, 400,000 acre-feet at 60 %, 70 % and 80 % and 600,000 acre-feet at 90 %.

Storage increments are added after the second routing sequence until one or no failures occur. Where the number of failure months is less than 75, increments equivalent to the initial storage assumption are applied. When the number of failures is greater than 75, the following increments are applied: one inch at 20 %, two inches at 40 % and 50 %; three inches at 60 %, four inches at 70 % and 80 %, and five inches at 90 %, in Case I; 100,000 acre-feet at 20 %, 200,000 acre-feet at 40 % and 50 %, 300,000 acre-feet at 60 %, 400,000 acre-feet at 70 % and 80 %, and 500,000 acre-feet at 90 %, in Case II; and 200,000 acre-feet at 20 %, 400,000 acre-feet at 40 % and 50 %, 600,000 acre-feet at 60 %, 800,000 acre-feet at 70 % and 80 %, and 1,000,000 acre feet at 90 %, in Case III.

The program provides for setting the initial condition of the hypothetical reservoir prior to each routing. It was assumed for these analyses that the reservoir would contain one-quarter of the assumed storages prior to testing each run.

In the program output shown in Figure 7-1, storages are shown in Column 1 in acre-feet. The figure in parentheses below the storage requirement at 1 or zero failures for each draft rate represents the storage surplus (negative) at zero failures or deficiency (positive) at 1 failure. The exact storage needed for a shortage index of zero can be obtained by subtracting the surplus or adding the deficiency to the final storage requirement of each draft rate. The data in column 2, in cfs, represent percentages of average flow from 20 % in the first set, 40 % in the second set, and increasing in increments of 10 % to 90 % of average flow. The long-term average of the historic record was used for this purpose since the model used to generate flow essentially preserves the mean flow.

The number of failures shown in Column 3 represents the total of failure months as shown in Columns 4 thru 15. For example, the 13 failures of two-month consecutive length equal 26 failure months. The average spillage, in Column 16, represents the amount of water wasted at the given storage requirement when the hypothetical reservoir is full. Column 18 gives the amount of water in storage at a selected month of the final year of the routing sequence. Columns 19 through 22 show the largest single monthly shortage and the average of all monthly shortages for each routing in both cfs and percent of the draft rate being routed as outflow.

In the processing of local areas where the flow from the upstream station or stations was subtracted from the flows at the downstream station, some negative flows resulted due to timing differences in observations. In cases where these negatives were obtained they were set to zero before starting the routing interval. The number of negative flows set equal to zero is indicated in the printouts for local areas.

FIGURE 7-1: SAMPLE OUTPUT OF STORAGE-YIELD MODEL (Source: SP-6)

*****/ANALYSIS/OF/YIELD/AT/STATION******
451 WEST BRANCH DELAWARE RIVER AT HALE EDDY,NY. DA 593

BASED ON 69 YEARS OF ACTUAL OR PARTIALLY ESTIMATED RECORD

STORAGE ACFT (DEFICIENCY)	AVG DRAFT CFS	NUMBER OF FAILURES	NUMBER OF FAILURES OF CONSECUTIVE MONTH LENGTH 1	2	3	4	5	6	7	8	9	10	11	12+	AVG ANNUAL SPILLAGE CFS	SHORTAGE INDEX	FINAL STORAGE ACFT	FAILURES 1ST IN CFS	AVG CFS	1ST IN PCT	AVG PCT
0	207	145	28	13	7	13	1	1	1	0	0	0	0	0	842	1.12183	0	202	89	90	41
15813	207	39	3	8	4	2	0	0	0	0	0	0	0	0	830	.21268	15813	176	88	82	42
31627	207	4	0	2	0	0	0	0	0	0	0	0	0	0	826	.01946	31627	155	86	80	44
47440	207	1	1	0	0	0	0	0	0	0	0	0	0	0	826	.00010	47440	21	21	11	11
(1251)																					
0	414	268	28	14	11	15	6	10	3	1	0	0	0	0	691	3.90178	0	430	217	95	51
31627	414	128	10	9	13	7	4	1	1	0	0	0	0	0	651	1.28405	31627	398	204	88	49
94880	414	10	3	0	1	1	0	0	0	0	0	0	0	0	621	.02903	94880	352	140	90	36
126507	414	3	1	1	0	0	0	0	0	0	0	0	0	0	619	.00010	126507	23	19	6	5
158133	414	0	0	0	0	0	0	0	0	0	0	0	0	0	618	0.00000	158133	0	0	0	0
(-29508)																					
0	517	314	28	17	14	9	14	10	4	2	0	0	0	0	627	5.54508	0	543	286	96	53
31627	517	176	9	12	9	14	8	1	2	0	0	0	0	0	578	2.50500	31627	532	293	94	55
94880	517	51	7	8	6	0	2	0	0	0	0	0	0	0	530	.28399	94880	455	223	93	45
126507	517	19	7	2	0	2	0	0	0	0	0	0	0	0	520	.04869	126507	450	155	92	32
158133	517	7	0	2	1	0	0	0	0	0	0	0	0	0	517	.01008	158133	214	127	44	26
189760	517	1	1	0	0	0	0	0	0	0	0	0	0	0	515	.00006	189760	41	41	9	9
(2501)																					
0	620	362	40	14	18	10	8	15	6	0	2	1	0	0	567	7.25494	0	657	348	97	54
63253	620	162	10	9	12	13	5	0	3	0	0	0	0	0	476	2.03940	63253	633	326	93	52
158133	620	30	8	4	0	2	0	1	0	0	0	0	0	0	421	.12944	158133	553	217	85	37
221387	620	10	0	2	0	0	0	1	0	0	0	0	0	0	415	.05614	221387	553	183	85	30
284640	620	5	0	0	0	0	1	0	0	0	0	0	0	0	412	.00556	284640	312	105	53	18
347893	620	0	0	0	0	0	0	0	0	0	0	0	0	0	411	0.00000	347893	0	0	0	0
(-31240)																					
0	724	404	39	18	16	14	10	14	9	0	2	1	0	0	516	9.04839	0	770	413	97	55
63253	724	208	17	13	12	12	12	0	3	0	0	0	0	0	413	3.31846	63253	746	412	94	56
189760	724	38	9	7	0	2	0	0	1	0	0	0	0	0	324	.26587	189760	712	308	90	45
253013	724	12	0	1	1	0	0	0	1	0	0	0	0	0	315	.15328	241305	712	335	90	47
316267	724	6	0	0	0	0	0	1	0	0	0	0	0	0	313	.07926	241305	664	365	84	50
379520	724	5	0	0	0	0	1	0	0	0	0	0	0	0	312	.01913	241305	410	230	60	33
442773	724	1	1	0	0	0	0	0	0	0	0	0	0	0	311	.00024	241305	112	112	17	17
(6804)																					
0	827	433	40	16	16	12	11	18	8	1	2	2	0	0	463	10.80498	0	884	488	98	57
63253	827	263	14	13	13	13	15	4	0	3	1	0	0	0	356	4.73442	63253	884	473	98	56
189760	827	79	5	13	8	0	2	1	0	1	0	0	0	0	244	.71895	189760	884	388	98	49
316267	827	20	0	3	2	0	0	0	0	1	0	0	0	0	217	.27894	196485	884	392	98	48
379520	827	12	0	1	1	0	0	0	1	0	0	0	0	0	214	.19497	196485	826	448	91	54
442773	827	11	0	1	1	0	0	1	0	0	0	0	0	0	213	.10880	196485	770	394	89	49
506027	827	10	0	1	1	0	1	0	0	0	0	0	0	0	213	.05449	196485	513	330	66	42
569280	827	8	0	1	2	0	0	0	0	0	0	0	0	0	212	.03204	196485	513	282	66	36
632533	827	4	0	2	0	0	0	0	0	0	0	0	0	0	211	.01299	196485	513	304	66	39
695787	827	1	1	0	0	0	0	0	0	0	0	0	0	0	210	.00046	238991	176	176	23	23
(10706)																					
0	930	469	41	18	12	12	12	19	10	2	2	3	0	0	417	12.57676	0	998	593	98	58

The program output was plotted as shown on Figure 7-2, showing storage requirements at different shortage indices for various demand rates. Storage values in Column 1 of the output (Figure 7-1) were converted to acre-feet per square mile and plotted against the corresponding shortage index in Column 17 of the output. The family of generated curves shown in Figure 7-2 is represented by the average of the 10 shortage index values (one for each of the ten 100 year generated periods) for each assumed storage and draft rate. This average was used to avoid placing particular emphasis on any 100 year trace, and also to avoid using what might be considered as the 1000 year drought, which would lead to more conservative results than necessary.

The curves based on the actual and partially estimated record are shown as dotted lines on Figure 7-2. The solid lines represent the average of the ten 100 year generated traces. At this station the severity of the 1960's drought is vividly indicated by the steep slope of the dotted lines for yields of 70 % and 80 % of the average flow at a shortage index of about 0.2. Because of the extended deficiencies during this drought, a substantial amount of storage would be necessary to provide enough annual carryover to maintain these yields through the latter years of the drought. For purposes of comparison, a separate

FIGURE 7-2 GRAPHIC DISPLAY OF STORAGE-YIELD MODEL OUTPUT
(Source: SP-10)

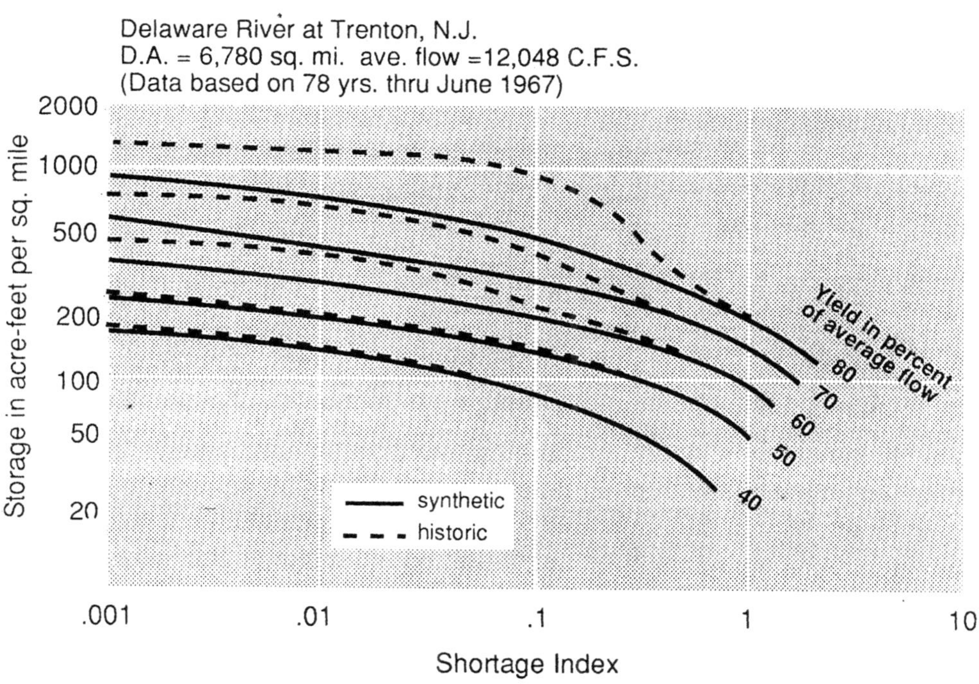

trial was made, using both the simulation and the yield-storage programs, in which the records of the latest drought were excluded. Figure 7-3 shows these results. In terms of the historic data, there is a marked reduction in storage requirements for a given yield and shortage index. However, it is interesting to note that the curves based on the synthetic records are relatively stable, even though 10 fresh sequences were generated and considerably fewer input data were used.

The relationships between historic and synthetic storage requirements shown in these two figures are not typical of all areas in the NAR. It was found that in many areas the historic data produced a smooth, stable family of curves. And on occasion a synthetic trace was found to contain a severe drought similar to that which affects the historic requirements in Figure 7-2. In many areas where the historic storage requirement was more critical than the generated flow requirement the drought of the 1960's was particularly severe. In 1964 and 1965, extreme conditions prevailed in southwestern New England, southeastern New York State, the Delaware River Basin, New Jersey, and the lower James River Basin.

FIGURE 7-3 GRAPHIC DISPLAY OF STORAGE-YIELD MODEL OUTPUT
(Source: SP-11)

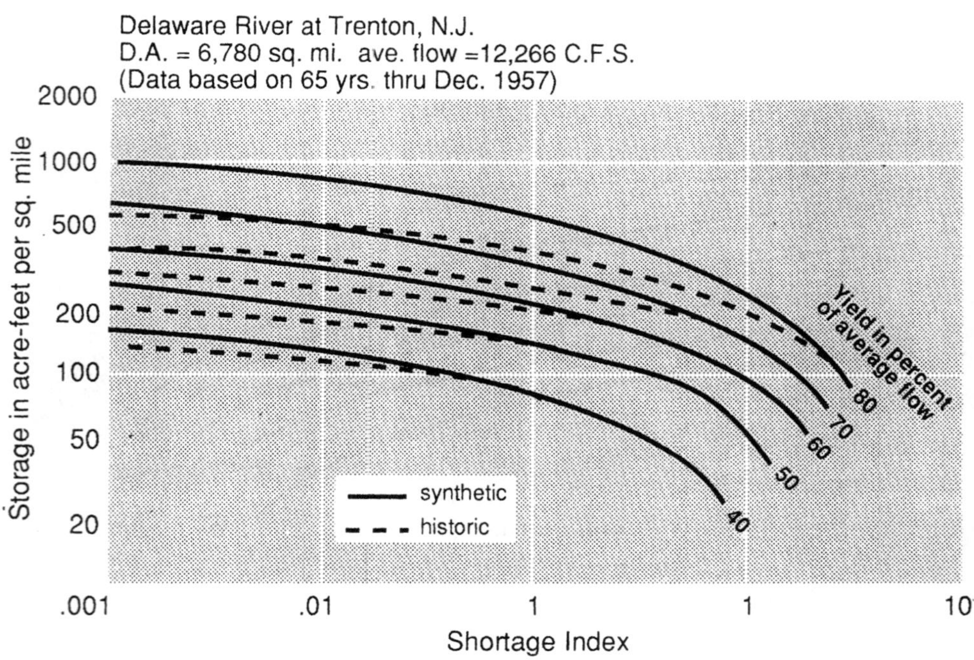

144 streamflow gaging stations were used as key stations for analysis in the NAR Study. These stations were selected to: provide suitable area coverage; use the longer-record stations available; and divide the major gaged portions of the region into subbasins that would be suited in size to the purposes of the study.

Storage requirements for the key stations in the NAR for selected shortage indices and percentages of average flow are shown in SP-12 to SP-17. For further discussion of results and comparative analyses, see Appendix C.

As previously stated, the storage at the beginning of each routing interval was set at 25% of the assumed storage (Column 1 of Figure 7-2). That is, the reservoir is one-quarter full when the routing period begins. This was assumed arbitrarily, since any condition could prevail at the beginning of operations at any particular site, and the timing and scope of analyses did not permit varying the assumption over a wide range of values. In several instances, large storage requirements resulted from critical drought conditions during the initial years of a generated streamflow trace. The average curves were manually smoothed in these cases under the assumption that full yield would not normally be required during the initial years of operation. It should be noted that initial volume assumptions can affect results and should be analyzed in the context of the specific requirements of detailed studies.

The shortage index used as the risk criterion for the model embodies both the number and the degree of severity of shortages; it is defined as equal to the sum of the squares of annual shortages during a 100 year period, where each annual shortage is expressed as a ratio to the annual requirement. It follows that the numerical value of the index varies according to the number of shortages and the square of the shortage quantities at a given demand. Thus in Table 7-1, one annual shortage of 10% in 100 years would be equivalent to an index of 0.01 and one shortage of 20 % in 100 years would be equivalent to an index of 0.04.

TABLE 7-1 EXAMPLES OF COMPUTATION OF SHORTAGE INDEX

ANNUAL SHORTAGES PER 100 YEARS	ANNUAL SHORTAGE PERCENTAGE	SHORTAGE INDEX
1	10	$(0.10)^2 = 0.01$
1	20	$(0.20)^2 = 0.04$
5	10	$5(0.10)^2 = 0.05$
10	20	$10(0.20)^2 = 0.40$

The shortage index concept implies that economic and social consequences of drought periods can be related to the square of the degree of shortage. Since the index reflects the magnitude of shortage as well as the number of shortages, it has considerable merit over frequency alone as a criterion, and it could possibly be multiplied by a constant to obtain an estimate of damages.

In actuality the computation of the shortage index would seldom be as simple as shown in Table 7-1. The index would most often represent a number of deficient years of widely varying shortage. For example, if in a 100 year period with a constant monthly demand of 1,000 cfs, there were a six month shortage of 400 cfs (2,400 cfs-months) in one year and a four month shortage of 300 cfs (1,200 cfs-months) in another year, the shortage index would be computed as follows:

$$\text{S.I.} = \left[\frac{2{,}400}{12 \times 1{,}000}\right]^2 + \left[\frac{1{,}200}{12 \times 1{,}000}\right]^2 = .04 + .01 = .05$$

Where there is a monthly variation in flow requirement, it is necessary to calculate monthly shortages, and sum and square for each year. If the period of analysis is less than 100 years, the calculated shortage index would be adjusted by multiplying it by the ratio of 100 divided by the period of record used.

Figure 7-4 further illustrates the relationships described in the definition of shortage index by depicting theoretical shortage versus number of shortage combinations for a wide range of indices. It should be noted that many combinations are possible. For example, although two 10% shortages would result in an index of 0.02, one 10% and two 7% shortages would also be approximately equivalent to an index of 0.02.

The additional scale in Figure 7-4 is useful for relating an equivalent three-month shortage to a given annual shortage. On the basis of a sampling of data from NAR analyses, three months has been found to be a fair estimate of the average length of shortage for comparative purposes. Obviously the maximum duration of shortage could be considerably longer depending on the requirements, the storage, if any, and the shortage index. The dashed parallelogram in Figure 7-4 serves to focus attention on the most likely areas and is based on data from the NAR yield-storage analyses for a wide variety of storage, yield and shortage index combinations. For example, a shortage index of 0.01 would generally involve less than four annual shortages.

7.3 PERSPECTIVES ON THE STORAGE-YIELD MODEL

There are well established engineering methods for the determination of the amount of storage needed for a specific amount of withdrawal

(Linsley and Franzini, 1964). The site-specific data required by these methods was not readily available for the NAR study, which covers a region of many rivers and potential storage sites. In addition, the limitations of the computational capabilities available at the time of the study had to be taken into account in developing the model. The storage-yield model described in this chapter is an attempt to deal with both of these factors. It produces generalized storage-yield

FIGURE 7-4 THEORETICAL RELATIONSHIP AMONG NUMBER OF SHORTAGES, SHORTAGES IN PERCENT, AND SHORTAGE INDEX (Source: SP-19)

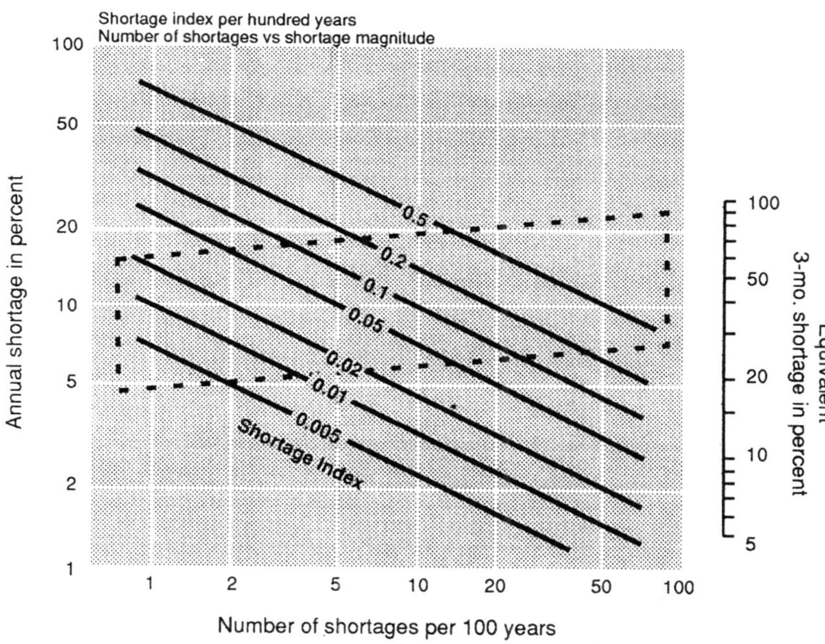

relationships for a large number of stations covering a large area and many rivers; the results of the model appear to be at a level of accuracy acceptable for framework planning.

From the computational standpoint, given present capabilities, it would be substantially easier now than it would have been at the time of the NAR study to develop improved regional storage-yield estimates through the use of simulation models for individual basins and the use of synthetic hydrology. The costs of site-specific data collection and organization will remain a factor in model choice for many regional studies. The choice of a generalized model, such as the model described here, versus a more detailed model for a particular study will depend on the extent to which the detailed model provides better estimates, the extent to which such estimates contribute to better decisions at the framework level, and the costs of developing the improved estimates.

REFERENCES

Fiering, Myron B and Barbara B. Jackson, Synthetic Streamflows, American Geophysical Union, Washington, D.C., 1971.

Linsley, Ray K., and J.B. Franzini, Water Resources Engineering, McGraw-Hill, New York, 1964.

Chapter 8: Institutional Arrangements

The water and related land resources planning study described in this book took place at a time of changing methods for water planning and within the context of a U.S. Water Resources Council planning authority that allowed more latitude than traditional agency planning procedures. This context provided freedom for the development of the institutional as well as the modeling aspects of planning; these institutional aspects are described here. The chapter includes a description of the institutional structure used for planning; a detailed presentation of the three stage planning process within which the development of the recommended plan took place; a discussion of the principal institutional methods used in the NAR study; a brief reference to traditional planning methods used in the study; and perspectives on the methods.

8.1 THE STRUCTURE FOR PLANNING IN THE NAR STUDY

The overall governing body of the NAR study was the Coordinating Committee, composed of one member from each of the participating Federal agencies, the NAR states, the District of Columbia, and the Delaware and New England River Basins Commissions. The chairman of the Coordinating Committee was the Division Engineer, North Atlantic Division, Corps of Engineers. The management structure for the study is given in Figure 8-1, which also lists the membership of the Coordinating Committee.

In the Terms of Reference published by the ad hoc Water Resources Council in June, 1965, the Department of the Army, Corps of Engineers, was assigned to chair the Coordinating Committee for the NAR (A-2, A-3). The Division Engineer of the North Atlantic Division (NAD) of the Corps of Engineers, New York City, was assigned the management function by the Office of the Chief of Engineers in April, 1966 (A-1). Within the North Atlantic Division office an NAR Study Group was established, which was later merged into the Special Studies Branch of the Planning Division. This group, together with similar groups in partner agencies, functioned as the staff of the Coordinating Committee. Day to day management activities were carried out by the Corps of Engineers acting as the Executive Agency. The head of the Special Studies Branch, acting as the Executive Secretary of the Coordinating Committee, was responsible for directing technical and administrative functions (A-17). The NAR Study Group performed study-related tasks as directed by the Division Engineer. At times special committees and task forces were employed, the most important of which was the Plan Formulation

FIGURE 8-1 MANAGEMENT STRUCTURE OF THE NAR STUDY (Source: A-4)

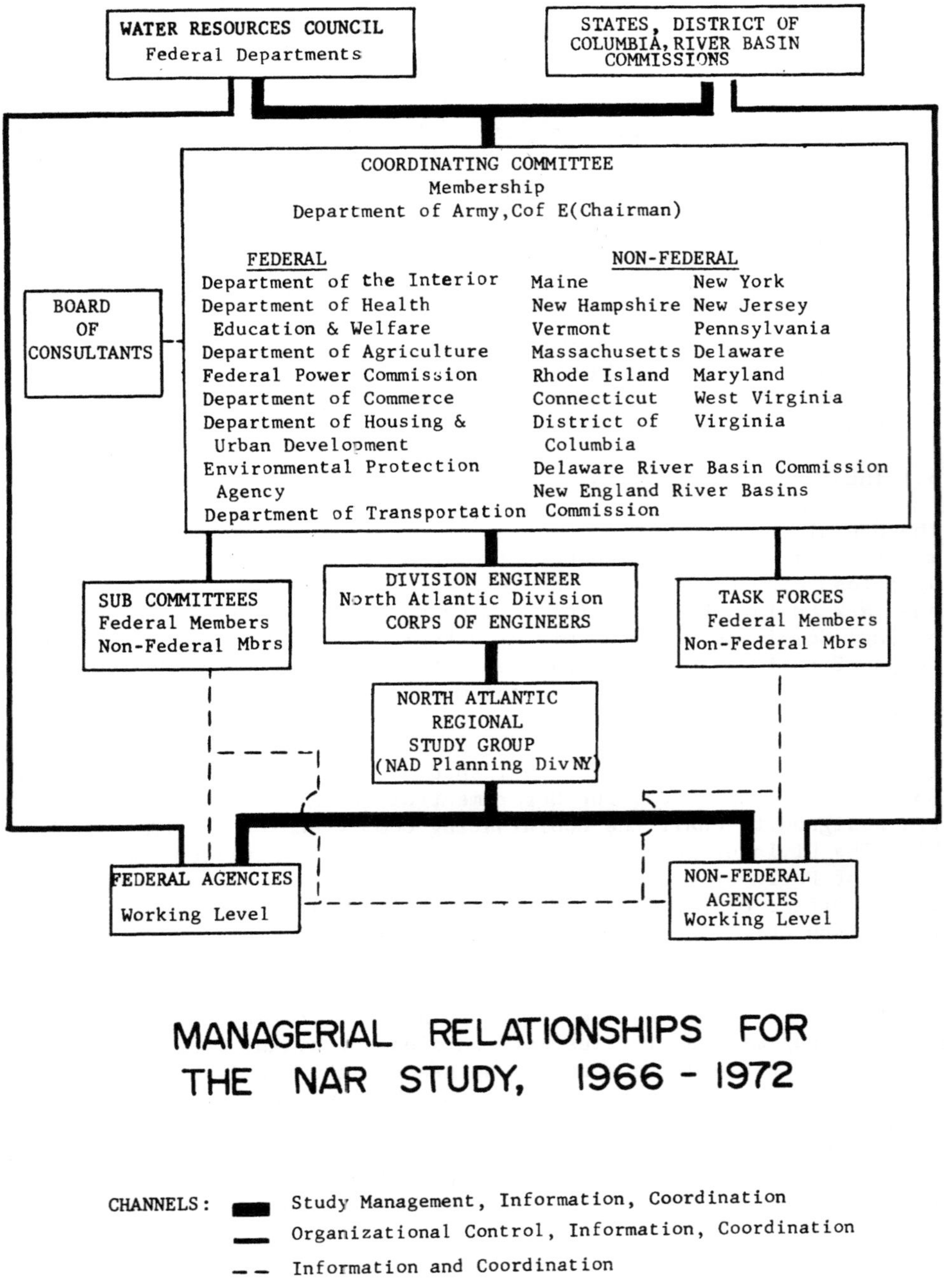

Work Group, which operated from 10 October, 1968, to the end of the study.

Despite the delegation of executive functions to the Corps of Engineers and other staff functions to various Federal agencies, the Coordinating Committee maintained active control of the study and final responsibility for its output. The Committee met 13 times from January, 1966 to April, 1972 and received numerous memos, drafts, and personal communications from the staff at and between these meetings. Most of the formal communication was in the form of Memoranda to the Coordinating Committee. There were a total of 164 of these between February 1966 and June 1972 (A-35); they transmitted both routine items and major work items.

An unusual feature of the institutional structure of the NAR, and one which some of the participants in the study rated as very important (Schwarz, Major, and Frost, 1975) to its success, was the Board of Consultants. The Board consisted of six eminent specialists in various disciplines related to water resources planning. The members of the Board served without compensation. The Board of Consultants was established and its members appointed at the second and third Coordinating Committee meetings, March, 1966 and June, 1966 (A-10).

The Board provided a broad professional overview of the study, pointing out shortcomings and suggesting new or modified approaches to various problems as the study progressed. It acted as a sounding board for ideas on the formulation of the study plan, reviewed the progress of the study, and brought the range of experience of its members to bear in recommending lines of approach. The members and their positions at the time of the study are:

Mr. David V. Auld	Consulting Engineer; Former Director of Sanitary Engineering, District of Columbia
Dr. Arthur Maass	Professor of Government, Harvard University
Mr. Eugene W. Weber	Consultant, Former Deputy Director, Civil Works, Corps of Engineers
Dr. Gilbert F. White	Professor of Geography, University of Colorado
Dr. Nathaniel Wollman	Dean, College of Arts and Sciences, University of New Mexico
Dr. Abel Wolman	Professor Emeritus, Johns Hopkins University

The Board held eight meetings between October 1966 and January 1971; the recommendations made at each are summarized on A-14 to A-16. Among the recommendations of the Board that influenced the study most are those relating to: the use of multiple objectives; the development of three alternative plans, one responsive to each objective; and the development and utilization of computer-based models for planning.

Several special committees and ad hoc groups were formed during the study (A-19, A-22, A-24). Of these, the most important from the standpoint of planning methods and procedures was the Plan Formulation Work Group. This group consisted of one or more representatives from each member of the Coordinating Committee: Federal agencies, States, Basin Commissions, and the District of Columbia. The establishment of the Work Group was proposed at the Coordinating Committee meeting of August, 1968, and approved at the meeting of November, 1968 (A-11). The importance of the group is indicated by the number of meetings it held: 21 in the period from October, 1968 to December, 1971. These meetings were held in different locations in the NAR in order to assure appropriate interaction among all study participants.

8.2 THE THREE STAGES OF PLANNING

The three-stage planning process within which the development of the final recommended plan took place is one of the most important institutional aspects of the NAR study. The three stages are described here. The stages took place, respectively, from September 1968 - June 1969; June 1969 - September 1970; and September 1970 - May 1972. The stages are described in T-1 to T-5 and in Schwarz, Major and Frost (1975).

While the stages were not completely separated in time in the sense that all of the activities of one stage were completed before the activities of the next began, the dates given provide a picture of the time required for each stage of plan formulation.

8.3 THE FIRST STAGE OF PLAN FORMULATION (September 1968 - June 1969)

Plan formulation in the NAR study began with preparations for and discussions at the fifth and sixth meetings of the Coordinating Committee held in September 1967 and May 1968. At these meetings the Committee approved the use of multiobjective planning for the NAR, including the use of the national income, regional development and environmental quality objectives for plan formulation. The Committee also approved the recommendation of the staff that three alternative sets of estimates of the planning elements (needs, sources, devices) should be developed. Each set was to represent desired levels of the planning elements for benchmark years, with a somewhat exaggerated emphasis on one of the planning objectives. These three sets of estimates were to provide the basis for the development of a recommended plan for the NAR.

Three papers written early in this stage of plan formulation remain accurate descriptions of the process. The papers are: "The Proposed Rationale for Plan Formulation", a description of the multiobjective approach; "The Structure of Plan Formulation", a description of the information about interrelationships among needs, sources and devices required for planning; and "Plan Formulation: Rationale, Structure and Procedures", a summary of the three proposed stages of plan formulation; all are in T, ch. 2.

In the first stage of plan formulation, estimates of needs, sources and devices for each of the three alternative programs were developed independently by the agencies responsible for studying different purposes. Multiobjective methods in planning were new and it was thought to be desirable to encourage all participants to explore alternative methods of contributing to the planning process. In addition, because of the absence of the large-scale computational models developed later, centralized formulation of alternative sets of estimates would have been difficult at this stage.

The Plan Formulation Work Group was established with representation from each of the member states and agencies of the Coordinating Committee. For both budget reasons and reasons of efficiency members of the Plan Formulation Work Group ordinarily attended only those meetings relevant to their technical or geographical areas although all members were free to attend all meetings of the group.

The first step in the Plan Formulation Work Group meetings for this stage was the presentation of estimates of needs, sources, and devices for water and related land management and development for each area, for each objective, and for each benchmark year. The estimates were presented by the Work Group members from the agency that had developed them. For convenience and consistency in organizing material, agency estimates were entered on standard information worksheets developed for the study. An example is in Figure 8-2.

These estimates were then discussed by all members of the Work Group and tentative agreements were reached on appropriate levels of the planning elements for each objective and benchmark year. Following each meeting the staff assembled the estimates agreed upon and made minor adjustments as appropriate. This procedure for first stage plan formulation was developed and tested in an experimental meeting before the series of actual plan formulation meetings began.

After the entire series of first stage Plan Formulation Work Group meetings was completed the information developed was assembled, modified as required by inquiries to the agencies, and reproduced for review. This information was published as the "First Report of Plan Formulation." This report includes descriptions of the assumptions underlying the projections for each need and summaries of the implications for planning areas of the programs emphasizing alternative objectives. Issued serially between May and August, 1969, this report consists of more than 500 pages of text and tables. An example of a data reporting sheet, developed like those for the initial agency

FIGURE 8-2 EXAMPLE OF WORKSHEET FOR FIRST SET OF PLAN FORMULATION DATA (Source: Report, 108)

Basin D-14

Year 1980

Emphasized Objective NI

Demand Wildlife (in thousands)

Quantities Available

		BIG GAME		SMALL GAME		WATER FOWL	
		Man days	$ Value	Man days	$ Value	Man days	$ Value
Physical Balance (single purpose)	Demand	371	1,484	1,150	2,121	88	352
	Capability	207	836	1,168	2,155	51	204
	Need	164		0		37	

Recommended Device (single purpose)

Total Benefits

Total Costs

Net Benefits

BIG GAME

The following devices should have been implemented under present bench[mark] year if big game needs are to be satisfied.

① Access and legislation. The legislation should allow maximum sustained yields.

② Maintain habitat at present levels.

If the above devices were implemented, the big game resources would support 584,000 man-days of hunting or enough to satisfy demand for 1980 benchmark year.

SMALL GAME

Overall there are no small game needs for 1980, but if small game needs are to be satisfied in future benchmark years, the following devices must be implemented at least by 1980:

① Access and legislation. The legislation should allow maximum sustained yield.

② Maintain habitat at least at 1980 levels.

WATERFOWL

The following devices should have been implemented under present bench[mark] year if waterfowl needs are to be satisfied.

① Access.

② Legislation (see explanation under present)

③ Maintain habitat at present level.

If these devices are implemented in present bench[mark] year the waterfowl resource will support 136,000 man-days of hunting.

estimates especially for the study, is in Figure 8-3. Excerpts of material from the "First Report on Plan Formulation" and related material are in T, ch. 3. The chapter includes Memorandum 69-28 to members of the Coordinating Committee, which introduced the report; an abstract of the report; a summary of the Work Group meetings; the material that was distributed for Area 9 (Southeastern New England) as an example of the material prepared in the first stage; and an evaluation and discussion of the plan formulation material.

One important lesson learned during the first stage of plan formulation was that adequate computational techniques would have to be developed for use in the remainder of the study. It became clear that the amount of information required for multiobjective analysis on the scale envisioned could not be successfully manipulated without the use of large computational models. Hence the development of these models occupied more and more staff time in the study. The requirements of sensitivity analysis, which appears to be a natural concomitant of multiobjective planning, also strongly suggested the need for computational models.

8.4 THE SECOND STAGE OF PLAN FORMULATION (June 1969 - September 1970)

The second stage of plan formulation centered about the development of the demand and supply models and the improvement and expansion of the data and projections developed in the first stage. Generally the work accomplished in the second stage was done without Plan Formulation Work Group meetings; instead, extensive contact was maintained between the planning staff in New York and the various cooperating agencies and states. Each agency assumed responsibility for the improvement and development of the material that it had presented during the first round of plan formulation. During this stage worksheets were introduced into the planning process that were designed to facilitate estimates of, and comments upon, the nature of interactions among planning and management devices and the positive and negative effects of these devices on needs and sources.

Revised planning materials were distributed for review as the product of the second stage of plan formulation. Chapter 4 of App. T includes the planning materials package for Area 9 (Southeastern New England) as it appeared at the end of the second stage, as well as an introductory memorandum (Memorandum 70-14) and a note on "Device Interactions." An example of a worksheet from the second stage of plan formulation is given in Figure 8-4. An example of a device interaction worksheet from the second stage of plan formulation is provided in Figure 8-5. These sheets were not designed as "final output to go with the recommendations of the study, but rather were utilized as references in discussions of work group and other meetings when consideration was given to combinations of methods and devices to be used in satisfying demands in the NAR areas" (T-136).

FIGURE 8-3 EXAMPLE OF WORKSHEET FOR SECOND SET OF PLAN FORMULATION DATA (Source: Report, 109)

NAR Area 14 - Summary Table 1

1980	2000	2020
1. Municipal and Industrial Water Supply		
Demand 780 mgd, 270 mgd icrement over present. Suggested devices: 53 mgd wells and river intakes. First cost: Surface water $0.0 Ground water $5.0 million Total $50. million	Demand 1090 mgd, 310 mgd increment over year 1980. Suggested devices: 58,400 AF storage in 5 reservoirs (111 mgd); 45 mgd wells; 111 mgd treatment plants (4 plants); river intakes. First cost: Surface watwer $129.0 million Ground water $ 4.0 million Total $133.0 million	Demand 1490 mgd, 400 mgd increment over year 2000. Suggested devices: 80,000 AF storage in 2 reservoirs (135 mgd); 300 mgd diversion from other basins; 135 mgd treatment plants (2 plants); river intakes. First Cost: Surface water: $311.0 million Ground water: 0.0 Total $311.0 million
2. & 5. Self-Supplied Industrial Process and Cooling Water		
Demand 730 mgd, 210 mgd increment over present. Suggested devices: Demand can be accommodated by presently available surface supply with intake structures and pumping facilities, at no cost for water supply.	Demand 1060 mgd, 330 mgd over year 1980 Suggested devices: Same as for 1980.	Demand 1560 mgd, 500 mgd increment over year 2000. Suggested devices: Same as for 1980.
3. Rural Domestic and Livestock Water Supply		
Demand 14 mgd, 1 mgd over present Suggested devices: 1 mgd wells. First cost: $0.1 million	Demand 17 mgd, 3 mgd over year 1980. Suggested devices: 3 mgd wells. First cost: $0.3 million	Demand 20 mgd, 3 mgd over year 2000. Suggested devices: 3 mgd wells. First cost: $0.3 million
4. Irrigation Water Supply		
Yearly demand 32,000 AF, 22,000 AF increment over present. Suggested devices: 9,000 AF storage and 13,000 Af wells and streams. First Cost: $5.7 million Annual cost: $0.8 million	Yearly demand 41,000 AF, 11,000 increment over year 1980. Suggested devices: 5,500 AF storage and 5,500 AF wells and streams First cost: $8.7 million Annual cost: $1.0 million	Yearly demand 53,000 AF, 14,000 increment over year 2000. Suggested devices: 8,400 AF storage and 5,600 AF wells and streams. First cost: $10.4 million Annual cost: $ 1.1 million

FIGURE 8-4: WORKSHEET FROM SECOND STAGE OF PLAN FORMULATION; DEMAND DATA SET (Source: Report, 110)

VISUAL AND CULTURAL ENVIRONMENT - DEMAND | POSSIBLE DEVICES | Area 14

OBJECTIVE	BENCH MARK YEAR	Unique Natural Landscapes Sq. Mi. and Shore Line in Miles	Unique Cultural Landscapes No. Sites	Quality Landscapes Sq. Mi. – Maintain	Quality Landscapes Sq. Mi. – Develop	Landscape Diversity Sq. Mi. – Maintain	Landscape Diversity Sq. Mi. – Develop	Maintain Ag. Landscape Sq. Mi.	Clean Water Percent of Area	Metropolitan Amenities-(Urban-Suburban) Percent of Area
	Pres		9							
EQ	1980		9		130	30			100%	2 M 5% 60 SM
	2000				130	30				
	2020				130	30				
NE	1980		9		65	15			100%	2 M 30 SM
	2000				65	15				30 SM
	2020				65	15				
RD	1980		9		130	30			100%	2 M 5% 60 SM
	2000				130	30				
	2020				130	30				

OBJECTIVE	BENCH MARK YEAR	Legal - Fee Simple Purchase (F.S.)	Legal; Purchase Leaseback (P.L.)	Legal - Easements (E.A.S.)	Legal - Deed Restrictions	Legal - Zoning	Legal - Tax Incentive-Subsidy (T.I.)	Impoundment	Water Supply Instal.	Waste Treatment	Recreation Facilities	Wildlife Facilities		Flood Plain Management – Preventive	Flood Plain Management – Corrective	Watershed Management – Ag. Practices	Watershed Management – Reforestation		Millions * COST
	Pres																		
EQ	1980	2M 205 SM		15 SM		X		X	X	X	X	X		X	X	X	X	FS Coast FS EAS	92.0 28.0 5.4
	2000	305 SM		15 SM				X	X		X	X		X	X	X	X	FS EAS	92.0 5.4
	2020	205 SM		15 SM				X	X		X	X		X	X	X	X	FS EAS	92.0 5.4
NE	1980	2M 30 SM	65 SM	15 SM		X		X	X	X	X	X		X	X	X	X	Coast FS FS PL EAS	13.5 [illegible] 29.0 (-2% Ann.) 5.4
	2000	30 SM	65 SM	15 SM				X	X		X	X		X	X	X	X	FS PL EAS	13.5 29.0 5.4
	2020		65 SM	15 SM				X	X		X	X		X	X	X	X	PL EAS	29.0 (-2% Ann.) 5.4
RD	1980	2M 205 SM		15 SM		X		X	X	X	X	X		X	X	X	X	FS Coast FS EAS	92.0 28.0 5.4
	2000	205 SM		15 SM				X	X		X	X		X	X	X	X	FS EAS	92.0 5.4
	2020	205 SM		15 SM				X	X		X	X		X	X	X	X	FS EAS	92.0 5.4

* Cost is given in 1970 values of land.

FIGURE 8-5: WORKSHEET USED IN SECOND STAGE OF PLAN FORMULATION; DEVICE INTERACTIONS (Source: Report, 111)

NON-MONETARY BENEFITS AND COSTS

AREA 14
OBJECTIVE All
NEED 5-Power Cooling

DEVICES	NEEDS / INTERACTIONS	BENEFITS EQ	BENEFITS NE	BENEFITS RD	EXC S-A
Wells Diversions Cooling Towers	1. Municipal and Industrial: Complementary use	O	O	O	
	2. Industrial Process and Cooling: Complementary use	M	M	M	
	3. Rural Water Supply:	O	O	O	
	4. Irrigation: High temperature water aids growth of some crops	O	O	O	
	6. Hydroelectric: Complementary use	O	O	O	
	7. Navigation: Complementary use	O	O	O	
	8. Recreation: More boating days if thermal pollution keeps streams from freezing	L	L	L	
	9. Fish and Wildlife: Possible improvement in crustacean and shellfish breeding in coastal areas	O	O	O	
	Increased use of area by resting waterfowl if kept from freezing	L	O	O	
	10. Solid Waste Disposal:	O	O	O	
	11. Liquid Waste Discharge: Higher water temperature increases assimilative capacity of stream	L	L	L	
	12. Flood Control:	O	O	O	
	13. Drainage Control: Withdrawal of groundwater lowers water table	O	O	O	
	14. Erosion Control:	O	O	O	
	15. Health: Warm water decreases sedimentary capacity of streams	O	O	O	

NEEDS / INTERACTIONS	COSTS [illegible]	COSTS [illegible]	COSTS [illegible]	EXC S-A
1. Municipal and Industrial: Higher water temperature could cause taste and odor problem	L	L	L	
Competitive use	O	M	O	
2. Industrial Process and Cooling: Taste and Odor	O	O	O	
3. Rural Water Supply: Competitive use of groundwater	O	O	O	
Higher temperature of return water	O	O	O	
4. Irrigation: Competitive use of water	O	O	O	
7. Navigation: Competitive water use	O	O	O	
8. Recreation: Competitive site use	O	O	O	
9. Fish and Wildlife: Thermal pollution disrupts natural habitat	L	H	L	
Harmful to marine organisms taken up into intakes	O	L	O	
Reduction of aquatic life by mortality through condensers	O	L	O	
10. Solid Waste Disposal:	O	O	O	
11. Liquid Waste Discharge: Water loss to stream	O	O	O	
Lowered D.O. levels resulting in increased sulfides and acidity	O	O	O	
12. Flood Control:	O	O	O	
13. Drainage Control:	O	O	O	
14. Erosion Control:	O	O	O	
15. Health: Fish kills	O	O	O	

Abbreviations

EXC. S-A = Except Sub-Areas
H = high
M = medium
L = low
O = zero or insig.

8.5 THE THIRD STAGE OF PLAN FORMULATION (September 1970 - May 1972)

This stage began with six Plan Formulation Work Group meetings, each held in or near one of the State capitals in the NAR. The purpose of these meetings was to formulate draft programs for a mixture of objectives for each of the NAR areas on the basis of the materials developed in stage two. The locations for the meetings were chosen to encourage the participation of large numbers of personnel from State and regional planning and water agencies. The series of work group meetings was preceded by meetings of NAR staff with personnel from each of the States; these meetings were for the purpose of familiarizing State personnel with the demand model outputs that were used during the work group meetings.

Following this series of meetings draft area programs were prepared by the NAR staff and agency staffs on the basis of agreements at the meetings. In the course of the third stage the program for each area was rewritten and reviewed several times as new information was obtained and comments were received.

The outputs from the demand model, which had first become available during the second stage of plan formulation, were fully incorporated into plan formulation during the third stage, beginning with the series of work group meetings. Outputs from the supply model took longer to obtain than had been anticipated and the third stage was extended to permit the full use of these results in plan formulation. A special work group meeting was held in 1972 to consider supply model results and to decide upon final runs reflecting mixed objectives. These runs were made by the NAR staff and the results were incorporated into the mixed objective area programs.

The members of the Coordinating Committee met during this stage sitting as the Plan Formulation Work Group. From the analysis of trade-offs made at this meeting emerged the mix of objectives on which the recommended NAR program is based.

The results of the third stage of plan formulation include the 21 Area Programs (Annex 1 to the NAR Report) and the Regional Program (Chapter 8 of the Report); these are the basis for the Findings and Recommendations of the Coordinating Committee (Chapters 9 and 10 of the Report). Other materials relating to the third stage of plan formulation, including the models, are found in detail in T, chs. 5-7. Chapter 9 of this book describes the study outputs.

8.6 THE MAIN INSTITUTIONAL ARRANGEMENTS IN THE NAR STUDY

The main institutional aspects of the NAR study can be seen from the description of the three stage planning process. The institutional methods that were used arose in large part, although not exclusively, because of the focus of the planning process on the goal of developing a mixed objective plan. (The direction of causation between goal and structure went both ways: the Board of Consultants, for example, helped to focus the study on this goal). The main institutional methods that were applied in the NAR include: the development of a specific planning structure, including not only the prescribed Coordinating Committee but also the Board of Consultants and the Plan Formulation Work Group; planning for substantial interaction and iteration involving the Coordinating Committee, the planning staff, and the staffs of the cooperating agencies (Major, 1977, pp. 56-57); the development of detailed reporting forms to be used by all participants throughout the process to insure that as far as possible information would be focussed on the needs of the study rather than presented in terms of unrelated agency procedures; and the use of the models. All of these contributed to the coherence and cohesiveness of the study.

No general detailed study has been done of the effects of each of the institutional methods referred to and their relative importance in the study. However, one staff member attempted to assess the impact on the planning process of one of the important elements, the models (Major, 1972). This assessment, from which the following material is drawn, deals first with the coordination and control functions of the models in the planning process.

The two principal models used in the NAR study, the demand and supply models, constituted a focus in the study that helped to channel agency efforts toward the main purposes of framework planning rather than toward such subsidiary tasks as preparing agency appendices. This impact on the planning process was of especial importance within the institutional context for framework planning, in which the central staff included fewer than 10 professionals and the cooperating agencies were substantially independent.

The models provided methods of coordinating and controlling the form, quantity, and quality of agency inputs. The use of models imposes requirements on the form and quantity of inputs that must be met if agency inputs are to be used effectively. In addition, the use of models tends to encourage careful statements of assumptions and the documentation of procedures for generating inputs; this work in turn opens up channels of inquiry for the central staff. Thus the central staff acquires a degree of control over agency contributions that is not foreseen in the framework plan organization chart.

Computational models help control the quality of agency inputs for another reason. The construction of a large-scale model tends to bring with it the generation of estimates for many components of the model by the central staff, either because of the nature of the model (the input-output system, for example, normally provides projections for all sectors of the economy) or because of the need to generate rough estimates for initial exploratory runs of the model (e.g., cost functions for mathematical programming supply models). In either case the construction of the models creates insights and perspectives into the quality of inputs required. The central staff thus acquires criteria with which to evaluate and hence to coordinate and control the inputs of the participating agencies.

The assessment of the role of models in the planning process in Major (1972) also concerns their effectiveness in dealing with the substantial information requirements of multiobjective planning and sensitivity analysis; in facilitating the exercise of judgment because of the relative ease of laying out alternatives; and in providing the opportunity for a continuous planning process.

This last idea, that of a continuous planning process and of the role of the methods and models in providing for it following the completion of the report, was of importance to the planners (Schwarz and Major, 1971), and thus may have increased the extent to which they made the NAR planning process an explicit one. In T-6, it is suggested that one use of the methods and models was to serve as a comprehensive basis for updating the NAR study; the models provide the basis for a continuous planning process in the NAR region. The adoption of such a process would be an important step forward in water resources planning, where frequently planning teams have been disbanded following the completion of a report without leaving behind them a set of methods and models that would permit plan reformulation in response to changing conditions.

In particular, the demand model and the supply model were among the principal reflections in the study of the idea of utilizing an explicit planning process subject to repetition in the future. For replanning the NAR, several uses of these models were foreseen (T-6). The demand model could be easily updated with the use of the most recent input-output table, reduced to a regional table by the methods used to regionalize the updated 1958 table used in the model (see the Appendix to Chapter 5 in this book). In addition, the model could be used in its present form with the alteration of certain output headings and the input of vectors of pollution loading coefficients to give estimates of the projected levels of various pollutants in the subbasins of the NAR. The supply model can be utilized with updated cost-yield functions and updated constraints reflecting regional and environmental objectives. Further, with some reformulation, the supply model could be converted into a net benefit maximizing model, a form with which further insight into the optimal configuration of management and investment devices for the NAR could be obtained.

Another factor that might have increased the explicitness of the NAR planning process, and its potential replicability, was the planners' concern with the possible use of the methods in other studies (T-5);

"The NAR plan formulation methods and models may be of use to other water resources planners both in demonstrating on an overall basis how one group of planners attempted to deal with the complex multiobjective nature of planning in a cooperative context and on the more particular basis of providing methods and tools that may be borrowed for other planning efforts."

Some detailed possibilities for the use of the models in other studies are also provided (T-6):

> "For other planning efforts, the demand model can serve as a water withdrawal or pollution loading model for other areas where input-output tables are available or can be derived from national coefficients; and, where such tables are not available, it can be used as a flexible projection model based on other types of projection mechanisms. The supply model can be utilized generally as formulated now for many other regions, with new input data and alternative formulations of the equations to suit alternative import-export possibilities."

8.7 TRADITIONAL WATER RESOURCE PLANNING METHODS

Throughout this chapter, emphasis has been on the innovative methods utilized in the NAR study. However, it should be observed once again that a substantial proportion of the study was based on traditional agency planning procedures developed over a period of many years. Some of the output of the study depended directly on such procedures; perhaps more of the output was the result of these procedures as they were modified and channeled by the models and institutional methods applied in the NAR study and described in this book. As an example of the latter effect, one can cite the development, specifically for the supply model, of the first generalized groundwater cost estimates produced by the United States Geological Survey (Appendix D and Cederstrom, 1973).

8.8 PERSPECTIVES ON INSTITUTIONAL ARRANGEMENTS FOR THE STUDY

The institutional arrangements in the NAR study were influenced by the particularities of the U.S. water planning process at the time of the study. Abstracting from these, the study management consisted of: 1) a Board of Directors (the Coordinating Committee); 2) a Chief Executive Officer (CEO) (the Division Engineer); 3) a Staff Director (who was both the Project Manager and Executive Secretary of the Coordinating Committee); 4) a project staff; and 5) an independent advisory board (the Board of Consultants). Viewed in this light, features of the organization are transferable to a variety of planning problems and governmental arrangements. Among the considerations important to the successful use of these elements in planning are: the breadth of membership on the Board of Directors (in the NAR, representatives from all of the participating States and agencies); the quality of the CEO, the Staff Director and his or her staff, including the staff's ability to work effectively with contributing agencies; and the competence and total independence of the advisory board.

The use of three stages or rounds of planning was based on the fundamental importance in planning of iteration between planners and decision makers, as well as on particular features of the NAR study such as the development over time of the methods and models employed. Large planning operations are so complex that it cannot be expected that a single round of effort will result in the development of alternatives adequate for consideration by decision makers. At least several stages are required, with the number depending on the complexity of the problem. Highlighting these stages at the beginning of a plan helps to focus the attention of both planners and decision makers on the iterative nature of the planning process.

The special forms and methods of communication, such as model executive routines, developed for the NAR study can in large part be considered as precursors of the products of the user-friendly computer hardware and software now available. For most current studies, the forms, executive routines, and other methods that were specially crafted for the NAR study can be developed and improved upon with relative ease using standard computer packages; and the interchange of information will be enormously facilitated by the digital transfer methods now widely available.

REFERENCES

Cederstrom, D. J., "Cost Analysis of Ground-Water Supplies in the North Atlantic Region, 1970," U.S. Geological Survey, Water-Supply Paper 2034, 1973.

Major, David C., "Impact of Systems Techniques on the Planning Process," Water Resources Research 8:3, June, 1972, 766-768.

Major, David C., Multiobjective Water Resource Planning, American Geophysical Union, Washington, D.C., 1977.

Schwarz, Harry E., and David C. Major, "An Experience in Planning: The Systems Approach," Water Spectrum 3:3, Fall, 1971, 29-34.

Schwarz Harry E., David C. Major and John E. Frost Jr., "The North Atlantic Regional Water Resources Study," in J. Ernest Flack, ed., Proceedings of the Conference on Interdisciplinary Analysis of Water Resources Systems, American Society of Civil Engineers, New York, 1975, 245-271.

PART III: OUTPUTS AND PERSPECTIVES

CHAPTER 9: OUTPUTS OF THE STUDY

9.1 INTRODUCTION

This chapter describes the outputs of the NAR study. These outputs are the result of the planning process described in this volume. The main characteristics of this process can be summarized as follows. First, there was the use of multiobjective theory as a conceptual framework to guide the organization and evaluation of information in the study (Chapters 3 and 4). Second, there was the use of the demand, supply and storage-yield models to organize a substantial proportion of the analysis of the study (Chapters 5, 6 and 7). Finally, there were the institutional arrangements described in Chapter 8. The planning process also incorporated standard agency methods described in Appendices to the Report.

The principal outputs of the study are contained in the Report, Annex I, and Annex II. The outputs in each are summarized here and are described in more detail below. The Report includes the recommended regional mixed objective program; perspectives on objectives for the 21 areas; and discussions of needs, devices, benefits and costs for the areas. Annex I contains the detailed recommended programs for each area, and Annex II contains the area information reorganized along State boundaries to form the recommended State plans.

9.2 OUTPUTS

The principal output contained in the Report is the recommended mixed objective plan for the NAR (Report, 167 and 179-202.) This is presented in summary form in terms of the water, land, and environmental management components of the plan, reproduced here as Tables 9-1, 9-2, 9-3, and 9-4 (Report, 180-181, 182, 192, 195). This regional mixed objective plan was developed from the mixed objective plans elaborated for each area. These latter were developed on the basis of agreed mixes of objectives for each area (Report, 161) and then developed through further Plan Formulation Work Group analysis (Report, 164).

The recommended mixed objective plan for the region includes the needs, sources, and devices that, in combination, were thought to fulfill in the best attainable way the desired mix of objectives for the region. The dimensions of the recommended plan can be indicated by reference to the tables. Publicly supplied water in the region is forecast to increase nearly 3 times over the planning period, from 5.5 billion gallons per day at the time of planning to 15.8 bgd in 2020 (Table 9-1.) This is to be supplied in large part by the addition of 3990

TABLE 9-1 MIXED OBJECTIVE REGIONAL PROGRAM - WATER MANAGEMENT
(Source: Report, 180-181)

NEEDS—cumulative		Pres.	1980	2000	2020
Publicly Supplied Water	(1000 mgd)	5.5	7.2	10.7	15.8
Industrial Self-Supplied Water	(1000 mgd)	3.9	7.2	13.1	21.8
Rural Water Supply	(mgd)	400	570	790	720
Irrigation Water: agriculture	(1000 afy)	150	540	730	760
non-agriculture	(1000 afy)	100	270	460	700
Hydroelectric Power Generation	(1000 mw)	5	14	42	102
Navigation: commercial	(m. tons annually)	600	760	1130	1790
recreational boating	(m. boats)	1.6	2.1	3.5	6.0
Power Plant Cooling:					
withdrawal, saline	(1000 cfs)	23	43	131	260
brackish	(1000 cfs)	12	29	42	45
fresh	(1000 cfs)	10	12	21	30
consumption, brackish	(cfs)	120	280	430	510
fresh	(cfs)	120	480	1180	2340
Water Quality Maint.: non-industrial	(m. PEs)	44	56	70	86
industrial	(m. PEs)	70	140	300	620

DEVICES—incremental		Purposes	1980	2000	2020
Storage Facilities:					
reservoirs, upstream	(1000 af)	FW * †	680	850	2260
mainstream	(1000 af)	FW,WQ * †	340	1240	2410
Withdrawal Facilities:					
intakes & pumping, fresh	(1000 mgd)	PS,Ind **	3.1	6.3	10.2
brackish	(1000 mgd)	Ind	2.6	4.5	6.9
estuarine		Pow	(7)	(11)	(11)
ocean		Pow	(11)	(14)	(14)
waste water	(mgd)	Ind	150	380	820
wells	(mgd)	*	660	1050	1220
Conveyance Facilities:					
inter-basin diversions, into	(mgd)	* ≠	500	1200	2300
out of	(mgd)	*	470	1200	2300
Quality Control Facilities:					
temperature, cooling towers & ponds		Pow,WQ,Rec	(14)	(18)	(21)
chemical/biological					
potable water treat. plants	(mgd)	PS	520	2200	3610
waste treatment plants					
secondary (85%)	(m. PE removed)	WQ,VC,Rec	55	0	0
secondary (90%)	(m. PE removed)	WQ,VC,Rec	120	340	640
advanced (95%)	(m. PE removed)	WQ,VC,Rec	6	17	35
effluent irrigation		WQ,VC,Irrig	(7)	(7)	(7)
nutrient control		WQ,VC,Rec	(20)	(20)	(20)
stormwater discharge control		WQ,VC,Rec	(18)	(14)	(14)
acid mine drainage control		WQ,VC	(3)	(3)	(3)
septic tank control		WQ,VC,Rec	(12)	(12)	(12)
separate combined sewers		WQ,VC,Rec	(20)	(15)	(15)

* From the supply model for the following purposes: PS, Ind, Rur, Irrig, Pow.
** Also includes Pow, Irrig.
† Also includes Rec, VC.
≠ Includes 25 mgd from outside the Region, in 1980 only.
( ) Indicates number of basins to which applicable.

TABLE 9-1 (CONT.)

DEVICES—incremental (cont.)	Purposes	1980	2000	2020
Pumped Storage	HPG	(7)	(11)	(13)
Desalting Facilities (mgd)	Ind *	10	200	210
Monitoring Facilities	WQ,VC,Rec	(11)	(11)	(11)
Waterway Management:				
navigation channel improvement	Nav	(13)	(14)	(8)
debris removal	VC	(2)	(2)	(2)
recreation boating facilities	Rec,Nav	(21)	(21)	(21)
FIRST COSTS—incremental ($ million 1970)				
Water Development Costs:				
storage, upstream ≠		180	170	500
mainstream ≠		170	360	700
wells ≠		210	210	140
desalting ≠		40	460	440
Water Withdrawal and Conveyance Costs:				
inter-basin transfers ≠		100	720	820
public water supply and treatment		380	1060	1430
industrial self-supplied water		30	54	83
irrigation, agriculture		62	54	17
non-agriculture		150	140	190
Navigation: commercial		920	1410	460
recreational		70	130	170
Power Plant Cooling Water		0	480	1360
Water Quality Maintenance:				
waste treatment, secondary		11000	25000	48000
advanced		950	3550	7310
comb. sewer overflow & acid mine drainage control		7500	0	0

* From the supply model for the following purposes: PS, Ind, Rur, Irrig, Pow.

≠ From the supply model and includes OMR costs.

thousand acre-feet of mainstream storage in the region by 2020, at a total first cost in 1970 dollars of $1.23 billion (Table 9-1). Flood plain management programs for upstream areas are to be implemented for 2,960,000 acres (Table 9-3); the maintenance of unique natural landscape by a variety of devices is to increase from 11,000 square miles at the time of planning to 26,000 square miles in 1980 (Table 9-4). As these figures illustrate, the recommended plan includes proposals for substantial programs in water, land, and environmental management. By far the largest expenditures shown are those for water quality maintenance programs, a total of $103,310 million in 1970 dollars for first costs by the year 2020.

The estimated federal and non-federal shares of the first costs in the recommended plan are shown in Table 9-5 (Report, 202). The shares shown are the planners' estimates of how program costs might be distributed taking into account both the then-existing cost sharing formulas and trends in those formulas (Report, 201).

TABLE 9-2 (Source: Report, 182)

MAJOR WATER DEVELOPMENT PROGRAM FOR NAR 1980 - 2020*
(million gallons per day in increments)

Area	Upstream Reservoirs 1980	2000	2020	Mainstream Reservoirs 1980	2000	2020	Groundwater 1980	2000	2020	Desalting 1980	2000	2020	Transfers OUT P	1980	2000	2020	To Area	Transfers IN 1980	2000	2020	From Area	Total Developed in Basin 1980	2000	2020	Total** Developed for Basin 1980	2000	2020	Area
1	24	16	0	27	49	120	17	17	26													68	82	146	68	82	146	1
2		13	0				2	3	1													2	16	1	2	16	1	2
3	3	14	0				6	15	5													9	29	5	9	29	5	3
4	5	4	0				4	7	3													9	11	3	9	11	3	4
5	5	11	55				5	4	5													10	15	60	10	15	60	5
6	31	66	103				9	11	5													40	77	108	40	77	108	6
7	1	2	457	24	769	367	3	7	6				290	118	486	743	9a,b	118	100	7	8c	28	778	830	28	298	197	7
8	21	227	0	123	67	370	28	13	-19				195	118	100	7	7b					172	307	351	54	13	344	8
9	149	0	0	30	0	0	30	10	8									118	486	743	7b	209	10	8	327	496	751	9
10	28	19	175	117	257	185	19	20	133			1										164	296	493	164	296	493	10
11	9	58	28		26	116	17	23	8													26	107	157	26	107	157	11
12	86	42	699	37	1272	757	45	55	4				1380	0	270	1015	13b, 14	0	0	0	15a	168	1369	1460	168	550	353	12
13							152	209	293									0	270	485	12c	152	209	293	152	479	778	13
14	140	0	0		150	0	9	119	16									35	300	530	15b, 12c	149	269	16	184	569	546	14
15	39	55	761	11	54	319	27	43	207				605	35	300	0	12c, 14	37	0	0	17e	76	152	1287	62	67	647	15
16	12	0	0			35			36			6										12	0	77	12	0	77	16
17	67	40	66	299	351	1146	89	230	32				90	199	38	530	15d, 18a					455	621	1244	256	493	715	17
18	13	209	0				18	150	0		113	86	40	0	0	0	19c	162	38	530	17e	31	472	86	193	510	616	18
19	18	14	2	199	412	580	55	-16	120									0	0	0	18a	272	410	702	272	410	702	19
20	1	3	0	85	43	15	7	13	11													93	59	26	93	59	26	20
21	7	1	0	22	249	225	10	51	262	12	85	119						25	0	0	Chow. River	51	386	606	76	386	606	21
Chowan River	25	0	0										25	25	0	0	21c					25	0	0				Chowan River
Total	684	794	2346	974	3699	4235	1172	1000	1181	12	198	212	2625	495	1194	2295		495	1194	2295		2842	5675	7954	2825	4963	7326	

* Contains all water development but does not consider effect of development on existing resources. The completion of the Tocks Island, Beltzville, Raystown and Gathright projects is assumed prior to 1980.

** Development for the basin includes the sums of ground water, minimum upstream storage for irrigation and the difference between the remaining storage and transfers out of the basin.

TABLE 9-3 MIXED OBJECTIVE REGIONAL PROGRAM - LAND MANAGEMENT
(Source: Report, 192)

NEEDS—cumulative		Pres.	1980	2000	2020
Flood Damage Reduction:					
avg. ann. damage, upstream (m. $)		55	82	145	275
mainstream (m. $)		80	130	260	530
tidal & hurricane (m. $)		61	96	181	359
Drainage Control: cropland (m. acres)		1.2	1.7	2.6	2.8
forest land (1000 acres)		0	4	142	556
wet land		(1)	(1)	(0)	(0)
Erosion Control: agriculture* (m. acres)		15	19	22	23
urban and other (m. acres)		8	11	15	19
stream bank (1000 mi.)		(21)	0.5	1.5	2.5
coastal shoreline (1000 mi.)		(12)	0.9	1.9	2.0
DEVICES—incremental	Purposes				
Flood Plain Management:					
upstream (1000 acres)	FDR ≠		840	1600	520
mainstream	FDR ≠		(21)	(21)	(21)
Local Flood Protection:					
ocean (projects)	FDR		11	6	0
river (projects)	FDR		150	150	50
flood control channels (mi.)	FDR		810	970	260
Watershed Management (m. acres)	FDR, Drn ≠		5.5	7.6	6.7
Erosion Protection: land treatment	Ern		(21)	(21)	(21)
coastal shoreline	Ern, Rec, VC		(12)	(12)	(12)
river shoreline	Ern		(21)	(21)	(21)
Drainage Practices	Drn, FW		(20)	(20)	(20)
Flood Control Storage:					
upstream (1000 af)	FDR		800	880	510
mainstream (1000 af)	FDR		430	1030	460
FIRST COSTS—incremental ($ million 1970)					
Flood Damage Reduction: upstream			150	150	80
mainstream			700	650	40
Drainage Control			35	59	28
Erosion Control			2030	2120	870

* Includes cropland, pasture, and forest.
≠ Also includes FW, Rec, VC.
( ) Indicates number of basins to which applicable.

9.3 OBJECTIVES

The Report contains, in addition to the mixed objective regional plan, perspectives on the choice of objectives for the 21 areas (and thus for the mixed objective regional plan), together with discussions of needs, devices, benefits and costs for the areas for water, land, and environmental management.

The objectives selected for emphasis in each area (that is, the extent to which the recommended plan for that area would be based on each of the three alternative plans) are shown in Table 9-6 (Report, 162). The objectives for each area were initially selected by the State members

TABLE 9-4 MIXED OBJECTIVE REGIONAL PROGRAM - ENVIRONMENTAL MANAGEMENT (Source: Report, 155)

NEEDS—cumulative		Pres.	1980	2000	2020
Water Recreation: visitor days	(m.)	(21)	930	1480	2440
Fish & Wildlife: sport fishing man-days	(m.)	100	110	140	180
hunting man-days	(m.)	38	43	54	66
nature study man-days	(m.)	60	70	88	109
Health: vector control and pollution control		(21)	(21)	(21)	(21)
Visual and Cultural:					
landscape maintenance					
unique natural	(1000 sq. mi.)	11	26	26	26
unique shoreline	(mi.)	90	1360	1360	1360
high quality	(1000 sq. mi.)	4	12	19	26
diversity	(1000 sq. mi.)	(10)	3.7	7.2	10.2
landscape development					
quality	(1000 sq. mi.)	(6)	0.9	1.9	2.8
metro. amenities	(mi.)	(1)	2	2	2
metro. amenities	(1000 sq. mi.)	(12)	0.66	0.67	0.67
DEVICES—incremental		**Purposes**			
Land Controls:					
fee simple purchase (buying)	(sq. mi.)	VC,FW,Rec	17200	2700	2600
fee simple purchase (buying)	(mi.)	VC,FW,Rec	990	0	0
purchase lease	(sq. mi.)	VC,FW,Rec	2840	360	350
easements	(sq. mi.)	VC,FW,Rec	3600	3300	3100
deed restrictions		VC,FW	(1)	(1)	(1)
tax incentive subsidy	(sq. mi.)	VC,FW	700	450	300
zoning	(sq. mi.)	VC,FW,Rec	4830	450	300
zoning	(mi.)	VC,FW,Rec	260	0	0
zoning and/or tax inc. subs.	(sq. mi.)	VC,FW,Rec	7700	3600	3200
zoning and/or tax inc. subs.	(mi.)	VC,FW	32	0	0
Facilities:					
recreation development		Rec	(21)	(21)	(21)
overland transportation to facility		Rec	(17)	(17)	(17)
parking and trails		FW,VC,Rec	(21)	(21)	(21)
site sanitation and utilities		VC,Rec	(21)	(19)	(19)
Biological:					
habitat management, fish		FW	(21)	(21)	(21)
wildlife		FW	(21)	(21)	(21)
fishways		FW	(12)	(12)	(12)
stocking, fish		FW	(16)	(16)	(16)
wildlife		FW	(19)	(19)	(19)
insect control		Hlth,Rec	(21)	(21)	(21)
FIRST COSTS—incremental ($ million 1970)					
Visual and Cultural			6500	1200	1200
Water Recreation			6200	4300	6300
Fish and Wildlife: fishing*			64	86	102

* Includes annual operations and maintenance costs.

( ) Indicates number of basins to which applicable.

TABLE 9-5 ESTIMATED FEDERAL/NON-FEDERAL COSTS AND COST SHARES FOR NAR PROGRAM, 1980-2020 (Source: Report, 202)

	FIRST COSTS ($ million 1970) FEDERAL			NON-FEDERAL			SHARES IN PERCENT	
	1980	2000	2020	1980	2000	2020	Federal	Non-Federal
PROGRAM FOR WATER MANAGEMENT								
Water Development Costs:								
storage, upstream	102	108	300	68	72	200	60	40
mainstream	114	222	432	76	148	288	60	40
wells	0	0	0	210	220	140	0	100
desalting	0	0	0	38	458	436	40	60
Water Distribution Costs:								
inter-basin transfers	9	253	177	14	379	265	40	60
public water supply	0	0	0	380	1,080	1,450	25	75
industrial self-supplied water	0	0	0	29	53	81	0	100
irrigation, agriculture	0	0	0	60	57	18	0	100
nonagriculture	0	0	0	150	140	190	0	100
Navigation: commercial	1,240	1,200	880	0	0	0	100	0
recreational	33.5	68	84	33.5	68	84	50	50
Power Plant Cooling Water	0	0	0	0	590	1,660	0	100
Water Quality Maint.								
waste treatment, secondary	6,050	13,750	26,400	4,950	11,250	21,600	55	45
advanced	545	1,975	3,658	445	1,615	2,992	55	45
other [1]	4,345	0	0	3,555	0	0	55	45
PROGRAM FOR LAND MANAGEMENT								
Flood Damage Reduction: upstream	128	112	90	42	38	30	75	25
mainstream	540	645	0	180	215	0	75	25
Drainage Control	14	24	12	21	36	17	40	60
Erosion Control	1,025	1,070	440	1,025	1,070	440	40	60
PROGRAM FOR ENVIRONMENTAL MANAGEMENT								
Visual and Cultural	0	0	0	6,500	1,200	1,200	0	100
Water Recreation	3,100	2,200	3,150	3,100	2,200	3,150	50	50
Fish and Wildlife: fishing	32	43	51	32	43	51	50	50

1. Combined sewer overflows control and acid mine drainage control.

TABLE 9-6 (Source: Report, 162)

MIXED OBJECTIVE FOR NAR AREAS

Area	Mixed Objective
1.	RD and EQ
2.	EQ and NI
3.	EQ and NI
4.	(a) EQ with some NI (b) NI and EQ
5.	EQ with some RD
6.	NI with some RD
7.	(a) EQ with some RD (b) RD with some EQ
8.	EQ and RD New Hampshire - EQ with some RD
9.	(a) EQ with some NI (a) RD with some EQ
10.	(a) NI and RD with some EQ
11.	Vermont - EQ with some RD New York - RD Adirondack Forest Preserve - EQ
12.	RD with EQ Adirondack Forest Preserve - EQ
13.	NI and EQ
14.	New York - RD with EQ New Jersey - EQ upstream NI downstream
15.	New York - RD with EQ Pennsylvania - EQ, RD with NI New Jersey - (portions of b) EQ (portions of d) NI, EQ and RD Delaware - (portions of d) EQ
16.	(a) NI with some EQ and RD (b) EQ with some RD and NI
17.	New York - RD and EQ Maryland - some EQ with NI Pennsylvania - RD and EQ
18.	Maryland - RD and EQ Delaware - EQ Virginia - RD, EQ for Barrier Islands
19.	NI with some EQ and RD Pennsylvania - EQ with some RD
20.	EQ with some NI
21.	RD with some EQ

of the Plan Formulation Work Group, and were then reviewed by the Coordinating Committee (Report, 161). As summarized in the Report, "Environmental Quality has been emphasized to the greatest extent throughout the region, followed by Regional Development and then National Income." (Report, 161.)

Table 9-7 (Report, 163) gives the "Implied Relative Values" in percentages for the objectives chosen in each area. These values are the approximate degrees to which the planners based the mixed objective plan for the area on each of the three alternative plans for that area (they are therefore not multiobjective weights). For example, for area 1, the St. John River basin, the recommended plan is based approximately 50% on the regional development alternative and 50% on the environmental quality alternative.

Table 9-8 (Report, 166) shows "Limitations on Objectives" for the 21 areas. These are considerations that became evident during plan formulation that, in the planners' views, suggested limitations on the objectives that could reasonably be chosen for some areas (Report,156).

TABLE 9-7 (Source: Report, 163)

IMPLIED RELATIVE VALUES OF MIXED OBJECTIVES FOR NAR AREAS

AREA	NAME	OBJECTIVE[1/] RD NI EQ
1.	St John River Basin	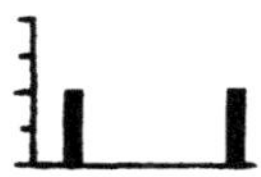
2.	Penobscot River Basin	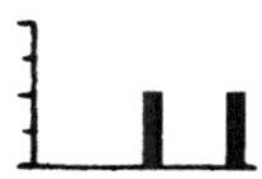
3.	Kennebec River Basin	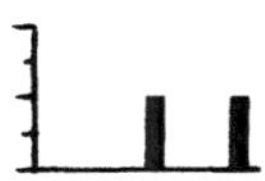
4.	Androscoggin River Basin	
5.	Maine Coastal Basins	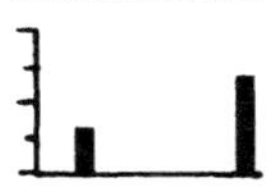
6.	Southern Me. and N.H. Coastal N.H.	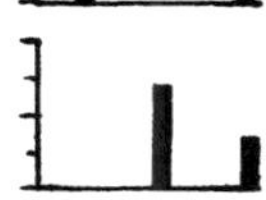
7.	Merrimack River Basin	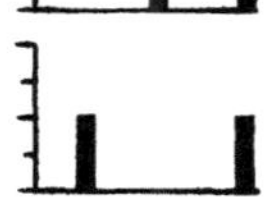
8.	Connecticut River Basin	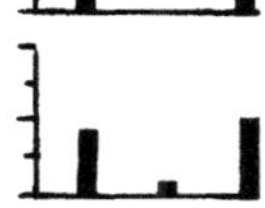
9.	Southeastern New England	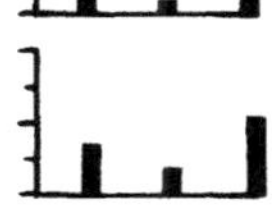
10.	Thames and Housatonic River Basins	
11.	Lake Champlain and St. Lawrence R.	
12.	Hudson River Basin	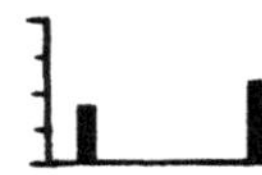
13.	Southeastern N.Y. Metropolitan Area	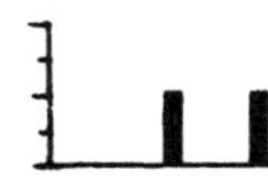
14.	Northern New Jersey	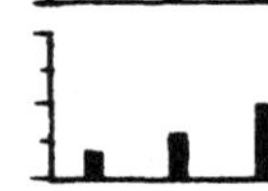
15.	Delaware River Basin	
16.	Coastal New Jersey	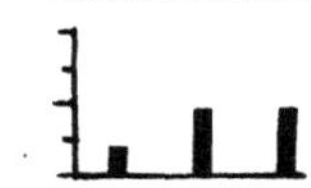
17.	Susquehanna River Basin	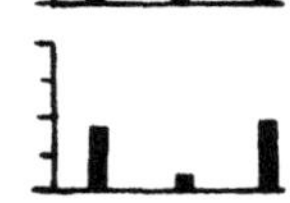
18.	Chesapeake Bay and Delmarva Peninsula	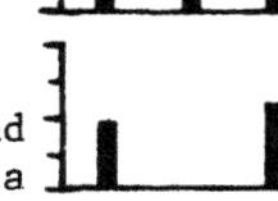
19.	Potomac River Basin	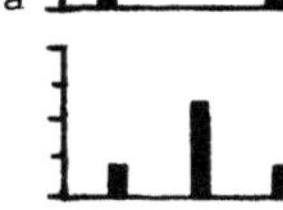
20.	York and Rappahannock	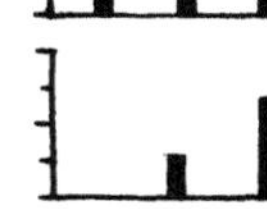
21.	James River Basin	

1/ Line represents percent of total of each objective as follows:

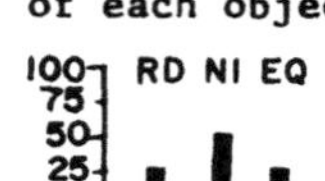

RD = 25%
NI = 50%
EQ = 25%

100%

TABLE 9-8 (Source: Report, 166)

LIMITATION ON OBJECTIVES
OF NAR AREAS

Area	Limitation	Area	Limitation
1.	None	11.	RD should receive some attention; no primary Emphasis upon NI alone; EQ alone in Adirondack Forest Preserve
2.	No objective emphasized alone and EQ should receive some attention	12.	EQ alone in Adirondack Forest
3.	EQ should receive some attention	13.	No primary emphasis on RD in Area as a whole exept for small pockets; EQ desired at some level but not alone,
4.	Different objective mixes for all sub-areas; RD will probably be limited due to lack of access	14.	None
5.	EQ should receive some attention	15.	Unlikely that any single objective should be emphasized alone
6.	EQ should receive some emphasis around Lake Sebago	16.	None
7.	None	17.	Objective mix should differ by subarea and by state due to diversity; EQ and RD should receive some minimal attention
8.	None		
9.	None	18.	None
10.	Different objective mixes for all sub-areas; no primary emphasis upon RD alone due to present high income	19.	None
		20.	None
		21.	None

Because of either the inadequacies or strengths of various resources, it was felt that "certain objectives could not be reasonably considered ...for primary emphasis or emphasis alone" (Report, 165) in ten areas. For example, it was held that only environmental quality should receive emphasis in the Adirondack Forest Preserve of areas 11 and 12 (Report, 165). It was judged that no such limitations would hold in the other 11 areas.

9.4 NEEDS, DEVICES, BENEFITS AND COSTS

The Report includes information on needs, devices, benefits and costs in the recommended plan for the 21 areas. This material expands upon the summary presentations of the recommended plan. It is intended to aid in fulfilling the purposes of framework planning "related to the identification of priorities for basin and project studies and to the identification of priorities for action within the Area programs." (Report, 165.)

For these purposes, information on needs, devices, benefits and costs for the areas is presented in tabular form according to a range of ranking criteria: important and "key" needs (the latter defined as needs that assist in filling other needs, Report, 118) for each area; areas ranked by size of each need, each need per capita, and growth rates of needs; important and key devices in each area; needs in each area likely to have large benefits when filled; device costs ranked within each area; ranking of areas by size of device costs and by size of device costs per capita.

The tables of important needs and devices are given here as Tables 9-9 and 9-10 (Report, 169, 174); the remaining nine tables are included in the Appendix to this chapter. In Tables 9-8 and 9-9 the important needs for Area 14, for example, the heavily developed northern New Jersey area, are given as publicly supplied water, recreational boating, water recreation, fish and wildlife, water quality maintenance, and flood damage reduction. The important devices are given as storage, withdrawal, conveyance (diversion) and water quality control (chemical/ biological) facilities.

The material in the tables is supplemented and enlarged by discussions of needs, devices, benefits and costs by area for each of the three components of the recommended plan: water, land, and environmental management (Report, 179-199). These discussions summarize the principal conclusions of the study relating to each of the elements.

With respect to the regional water management plan, for example (Table 9-1), the most important needs are given as publicly supplied water, industrial self-supplied water, power plant cooling, and water quality maintenance; and the areas and benchmark years in which these needs will be important are listed (Report, 179-184). With respect to water management devices, it is said that mainstream reservoirs will be the primary new sources of water in the region, and the areas and time periods in which reservoirs will and will not be significant sources are given (Report, 185). In the benefit section of the water management program summary it is reported that benefits from water quality maintenance should be the largest of those for any need in the region and the areas in which these benefits will be most important are given (Report, 189). Benefits were not in general estimated numerically in the study; in the summaries, they are described in terms of expected

TABLE 9-9 (Source: Report, 169)

TABLE 38
IMPORTANT NEEDS IN EACH NAR AREA

	Area	Publicly Supplied Water	Industrial Self-Supplied Water	Rural Water Supply	Irrigation Water	Power Plant Cooling	Hydroelectric Power Generation	Commercial Navigation	Recreational Boating	Water Recreation	Fish and Wildlife	Water Quality Maintenance	Flood Damage Reduction	Drainage Control	Erosion Control	Health	Visual and Cultural
1	ST. JOHN RIVER BASIN	X	X		X				X	X	X						
2	PENOBSCOT RIVER BASIN		X							X		X					X
3	KENNEBEC RIVER BASIN	X	X														X
4	ANDROSCOGGIN RIVER BASIN	X	X			X				X	X	X	X				X
5	MAINE COASTAL BASINS	X	X			X				X	X						X
6	SOUTHERN MAINE AND COASTAL NEW HAMPSHIRE	X	X	X		X		X									
7	MERRIMACK RIVER BASIN									X	X	X	X				X
8	CONNECTICUT RIVER BASIN									X	X	X					X
9	SOUTHEASTERN NEW ENGLAND	X	X			X			X	X		X				X	
10	THAMES AND HOUSATONIC RIVER BASINS	X	X							X		X					
11	LAKE CHAMPLAIN AND ST. LAWRENCE RIVER DRAINAGE	X	X														X
12	HUDSON RIVER BASIN	X	X									X	X				
13	SOUTHEASTERN NEW YORK METROPOLITAN AREA	X							X	X		X					X
14	NORTHERN NEW JERSEY	X							X	X	X	X	X				
15	DELAWARE RIVER BASIN	X	X					X		X	X	X	X				
16	COASTAL NEW JERSEY	X	X									X					
17	SUSQUEHANNA RIVER BASIN	X	X							X		X					
18	CHESAPEAKE BAY AND DELMARVA PENINSULA DRAINAGE											X					X
19	POTOMAC RIVER BASIN	X										X			X		X
20	RAPPAHANNOCK AND YORK RIVER BASINS	X	X	X						X	X						
21	JAMES RIVER BASIN		X			X		X									

TABLE 9-10 (Source: Report, 174)

TABLE 43
IMPORTANT DEVICES IN EACH NAR AREA

	Storage Facilities	Withdrawal Facilities	Wells	Conveyance Facilities (Diversions)	Quality Control, Temperature (Power Plant Cooling)	Quality Control, Chemical/Biological	Monitoring Facilities	Flood Plain Management	Local Flood Protection	Watershed Management	Erosion Protection	Drainage Practices	Waterway Management	Recreational Boating Facilities	Land Controls	Land Facilities	Habitat Management	Fishways	Stocking	Insect Control	Research	Education	Water Demand and Allocation Changes	Project Operation Changes
1 ST. JOHN RIVER BASIN						X									X					X				
2 PENOBSCOT RIVER BASIN															X									
3 KENNEBEC RIVER BASIN						X																		
4 ANDROSCOGGIN RIVER BASIN						X									X									
5 MAINE COASTAL BASINS																								
6 SOUTHERN MAINE AND COASTAL NEW HAMPSHIRE																								
7 MERRIMACK RIVER BASIN		X				X																		
8 CONNECTICUT RIVER BASIN	X			X	X	X																		
9 SOUTHEASTERN NEW ENGLAND				X		X				X					X		X						X	
10 THAMES AND HOUSATONIC RIVER BASINS	X					X																		
11 LAKE CHAMPLAIN AND ST. LAWRENCE RIVER DRAINAGE						X				X						X	X							
12 HUDSON RIVER BASIN	X	X				X																		
13 SOUTHEASTERN NEW YORK METROPOLITAN AREA				X		X																		
14 NORTHERN NEW JERSEY	X	X		X		X																		
15 DELAWARE RIVER BASIN	X	X	X		X	X																		
16 COASTAL NEW JERSEY																								
17 SUSQUEHANNA RIVER BASIN	X	X			X	X																		
18 CHESAPEAKE BAY AND DELMARVA PENINSULA DRAINAGE						X									X									
19 POTOMAC RIVER BASIN						X																		
20 RAPPAHANNOCK AND YORK RIVER BASINS																								
21 JAMES RIVER BASIN																								

magnitude, with "benefit" referring to positive impacts on the objectives (Annex 1, 11-12). Costs for the water management plan, which were as a rule estimated numerically for the national income objective, are described in general terms for devices and areas on Report, 190, where many devices are said to have large costs in at least some areas. Similar summaries for needs, devices, benefits and costs are given for the land and environmental management programs shown in Tables 9-3 and 9-4 (Report, 191-199).

9.5 AREA PROGRAMS

Annex I to the Report contains the 21 area programs. These detailed programs played a "central role in the planning activities of the NAR Study. It became clear during Plan Formulation that detailed descriptions were required of the choices made. These 21 descriptions of choices allow planners to easily and fully reevaluate their decisions and allow the public to evaluate the recommendations" (Annex I, 1). The format of the area programs is described in this section. The 18 page program and discussion for Area 9, Southeastern New England, is included in the Appendix to this chapter for readers who wish to examine study outputs in detail.

The presentation in Annex I of each area program begins with an outline map of the area and a description of those factors required to provide a "succinct impression of the outstanding characteristics which affect decisions about the Area's water resources management programs" (Annex I, 3). Next, the planning objectives that might reasonably be emphasized in the area are discussed, together with the considerations that might lead to particular choices among these. The mix of objectives selected for the area is identified, along with the reasoning that led to this choice. Then, need levels and their relations to objectives are discussed, followed by a similar discussion of devices. Benefits and costs are discussed, in terms of their general magnitudes, for different needs and devices. Next, the three alternative programs are compared to the selected mixed objective program. Finally, the mixed objective program and the three alternative programs are presented in tabular form, including needs, devices, and estimated first costs.

The material presented in Annex I for the mixed objective area programs was reformulated according to State boundaries and is presented for each State in Annex II. The purpose of the reformulation is to provide each State with a suitable summary of study results directly pertaining to it. This reformulation was done in such a manner as to insure that there would be "no changes in the plan formulation decisions on alternative planning elements: objectives, needs, devices, benefits and costs." (Annex II, 1.) The detailed methodology for this reformulation, including the percentages of each area allocated to each State, is given in Annex II, 3-26.

9.6 FINDINGS, CONCLUSIONS, AND RECOMMENDATIONS OF THE COORDINATING COMMITTEE

The final chapters in the Report present the findings, conclusions, and recommendations of the Coordinating Committee. These are based on the study output described in this chapter.

The most general finding of the Committee is that, while water and related land resources will on the whole be adequate in the region, development of these will become increasingly expensive in multiobjective terms. Need reduction programs will therefore become increasingly essential to the region. In the words of the Report, "...if the resource management programs outlined in this Study are generally followed and if the assumptions underlying the Study are proven accurate, the water and related land resources in the North Atlantic Region will be adequate to provide the services required by the Region's people through the year 2020 and beyond." (Report, 204.)

However, "large increases in the services supplied by water resources and required by the Region's people ...cannot be continuously fulfilled by the customary means of increasing natural resource consumption and use. Provision of services solely through development of natural resources must cease because the magnitudes and rates of increase in needs have become very large; the practical limits of resource development are being approached; the monetary and natural resource investments to meet these needs have become excessive; and the quality of life has become impaired. It is imperative that research, studies and programs be initiated now to find and implement means of reducing the needs for water and related land resources. This reduction of needs must occur no matter which objective is being sought although the degree of need control versus resource use does vary to some extent according to the objective." (Report, 204.)

The principal recommendation of the Committee, based on the study results, is that the 21 area plans should be adopted as framework plans for use in the NAR to "serve as guides and starting points for more detailed river basin or project planning" (Report, 220). This is qualified by the recommendation that "where more advanced planning has been accomplished (such as in the Delaware and Susquehanna Basins) these more detailed plans should serve as the guides" (Report, 220). There are other recommendations for studies and planning activities, both general and specific. For example, research programs relating to need reduction are recommended (Report, 220); and a detailed study of water and related land resources is urged for the Northern New Jersey area (Area 14), to begin almost immediately after the completion of the NAR (Report, 221). As to the NAR study itself, the Committee suggests an updating at 10 year intervals, the first updating to take place after the completion of the 1980 census (Report, 223).

A summary of the recommended program is provided (Report, 205-212), and the Committee details some of the numerous conflicts that might occur in the region over water and related land resources as a result of the program. Among these are conflicts relating to the multiobjective benefits and costs of the large increments of storage recommended for the region (Report, 215), pumped storage siting (Report, 216-17), and interbasin transfers (Report, 217).

The Committee also provides helpful perspectives on the limitations of the methods used as they were applied in the study. One important limitation is the "scarcity of information for all types of benefits and for non-monetary costs" (Report, 214), a scarcity that limits the ability of the planner to estimate trade-offs properly. (This is one of several data problems cited which will be familiar to water resources planners.) A second limitation discussed is the lack of variation by objective of some of the needs forecasts. Some of this reflects legal and technological constraints, exemplified by, respectively, legal standards for water quality and the limited number of sites available for hydroelectric power generation in the region. In such cases, according to the Committee, "these needs with single or limited levels were considered by the responsible agencies to reflect an already appropriate mixed objective. The standards for water quality, for instance, have already gone through a lengthy review process" (Report, 214). On the other hand, lack of variation by objectives for some needs represents lack of knowledge, as for example with respect to fish and wildlife needs (Report, 214). A third limitation cited is an insufficient number of iterations with the computer models. Additional runs would have permitted more review by the Plan Formulation Work Group of alternatives, and thus more refined choice among these (Report, 215). (This applies particularly to the supply model, which was developed later in the study than the demand model. The demand model was used for a large number of runs, reported on T-274 through T-279.)

A general problem relating to the implementation of multiobjective recommendations at the time of the study is cited by the Committee. According to the Committee, "National Income will be the most easily obtained objective in the NAR under the present procedures for water resources management. There will not be adequate and equal recognition of all objectives during program planning and implementation until formulation and evaluation procedures that are impartial for all objectives have been generally agreed on....The proposed new principles and standards of the Water Resources Council would fill part of this void" (Report, 212). (The principles and standards referred to are those of U.S. Water Resources Council, 1973, approved in the year following the submission of the NAR study.)

REFERENCE

United States Water Resources Council, "Water and Related Land Resources: Establishment of Principles and Standards for Planning," Federal Register 38:174, 1973, 24778-24869.

Appendix to Chapter 9

This appendix has two sections. The first, Area Tables: Rankings of Needs, Devices, Benefits and Costs, contains 9 tables (Report, 168, 170-173, 175-178) presenting information on needs, devices, benefits and costs for the 21 NAR areas according to ranking criteria given in the tables and summarized in Chapter 9. Together with Tables 9-9 and 9-10 these tables provide an overview of needs, devices, benefits and costs in the recommended plan for each area.

The second section of the appendix, Area Program for Area 9, Southeastern New England, contains the complete recommended mixed objective program for that area. The program for Area 9 is reproduced from Annex I, 142-159; the role of the area programs in the planning process is described in Chapter 9 of this book.

TABLE 37
KEY NEEDS IN EACH NAR AREA

		Publicly Supplied Water	Industrial Self-Supplied Water	Rural Water Supply	Irrigation Water	Power Plant Cooling	Hydroelectric Power Generation	Commercial Navigation	Recreational Boating	Water Recreation	Fish and Wildlife	Water Quality Maintenance	Flood Damage Reduction	Drainage Control	Erosion Control	Health	Visual and Cultural
1	ST. JOHN RIVER BASIN											X					X
2	PENOBSCOT RIVER BASIN											X					
3	KENNEBEC RIVER BASIN											X					
4	ANDROSCOGGIN RIVER BASIN											X					X
5	MAINE COASTAL BASINS											X					
6	SOUTHERN MAINE AND COASTAL NEW HAMPSHIRE																
7	MERRIMACK RIVER BASIN											X					
8	CONNECTICUT RIVER BASIN											X					
9	SOUTHEASTERN NEW ENGLAND										X	X			X		
10	THAMES AND HOUSATONIC RIVER BASINS																
11	LAKE CHAMPLAIN AND ST. LAWRENCE RIVER DRAINAGE																
12	HUDSON RIVER BASIN											X					
13	SOUTHEASTERN NEW YORK METROPOLITAN AREA	X										X					X
14	NORTHERN NEW JERSEY																X
15	DELAWARE RIVER BASIN								X			X					
16	COASTAL NEW JERSEY											X					
17	SUSQUEHANNA RIVER BASIN											X					
18	CHESAPEAKE BAY AND DELMARVA PENINSULA DRAINAGE											X					X
19	POTOMAC RIVER BASIN											X					X
20	RAPPAHANNOCK AND YORK RIVER BASINS																
21	JAMES RIVER BASIN											X					

TABLE 39
RANKING OF NAR AREAS
BY SIZES OF THEIR NEEDS

The rank shown is the highest that the Area achieved for that need during any planning period, including 1980, 2000 and 2020, and the present.

		Publicly Supplied Water	Industrial Self-Supplied Water	Rural Water Supply	Agricultural Irrigation Water	Power Plant Cooling, Fresh Water Withdrawal	Power Plant Cooling, Fresh Water Consumption	Hydroelectric Power Generation	Commercial Navigation	Recreational Boating	Water Recreation Visitor Days	Sport Fishing Man-Days	Hunting Man-Days	Nature Study Man-Days	Water Quality Maintenance	Flood Damage Reduction, Upstream	Flood Damage Reduction, Mainstream	Flood Damage Reduction, Tidal and Hurricane	Drainage, Cropland	Erosion, Agriculture	Erosion, Urban	Erosion, Streambank	Visual and Cultural, Maintenance of Unique Natural Areas	Visual and Cultural, Maintenance of Unique Shorelines	Visual and Cultural, Maintenance of High Quality Areas	Visual and Cultural, Maintenance of Diverse Areas	Visual and Cultural, Development of Metropolitan Amenities	Visual and Cultural, Development of Quality Areas
1	ST. JOHN RIVER BASIN	21	18	16	5	7	11	10	19	20	21	20	21	21	11	19	19	11	11	10	21	19	1	12	10	11	13	7
2	PENOBSCOT RIVER BASIN	18	8	20	16	8	8	9	12	17	19	17	16	18	6	18	18	11	16	19	18	16	5	12	1	11	13	7
3	KENNEBEC RIVER BASIN	17	17	18	11	7	15	6	18	16	18	19	19	19	14	15	15	11	10	14	19	15	18	12	2	11	13	7
4	ANDROSCOGGIN RIVER BASIN	18	13	21	12	10	15	7	19	19	20	18	20	17	12	15	12	11	17	21	20	16	9	12	4	11	13	7
5	MAINE COASTAL BASINS	19	14	15	17	18	18	13	14	16	17	16	18	17	15	17	20	10	13	12	11	19	11	1	4	11	13	7
6	SOUTHERN MAINE AND COASTAL NEW HAMPSHIRE	14	16	14	14	17	17	11	7	12	13	8	11	15	19	11	15	9	16	14	15	16	4	8	6	11	13	7
7	MERRIMACK RIVER BASIN	12	13	15	11	6	6	10	19	11	9	13	13	13	16	10	8	11	18	18	11	14	7	12	8	11	9	6
8	CONNECTICUT RIVER BASIN	10	4	9	8	5	4	3	11	7	8	9	6	9	7	6	6	11	7	9	5	9	4	12	6	11	13	7
9	SOUTHEASTERN NEW ENGLAND	3	8	13	5	19	19	16	6	2	3	2	7	3	4	2	11	3	19	17	6	12	9	10	10	11	5	2
10	THAMES AND HOUSATONIC RIVER BASINS	7	10	6	11	12	8	3	9	3	3	11	15	7	14	5	10	2	10	12	9	12	18	4	10	10	5	1
11	LAKE CHAMPLAIN AND ST. LAWRENCE RIVER DRAINAGE	14	10	9	6	9	7	2	14	14	14	10	5	14	18	14	7	11	3	7	15	6	2	12	9	7	13	7
12	HUDSON RIVER BASIN	7	4	4	4	4	3	1	8	6	7	14	10	9	5	5	4	11	5	3	3	8	1	12	1	2	5	7
13	SOUTHEASTERN NEW YORK METROPOLITAN AREA	1	19	7	2	19	19	19	4	1	2	1	11	1	1	21	14	1	21	18	4	19	18	4	10	7	3	7
14	NORTHERN NEW JERSEY	3	6	10	10	15	11	11	1	4	4	15	8	4	3	12	1	11	10	11	6	5	18	12	10	9	3	4
15	DELAWARE RIVER BASIN	2	1	2	1	2	1	3	1	3	1	3	1	2	1	6	3	6	4	5	2	3	10	11	10	1	1	7
16	COASTAL NEW JERSEY	12	21	17	3	19	19	19	16	14	12	6	16	12	14	19	21	5	10	16	13	10	14	9	10	11	11	5
17	SUSQUEHANNA RIVER BASIN	6	2	1	1	1	1	1	17	10	3	4	1	5	6	3	2	11	2	1	1	1	3	12	10	4	2	7
18	CHESAPEAKE BAY AND DELMARVA PENINSULA DRAINAGE	7	7	4	3	14	10	18	4	6	15	7	6	7	10	1	13	4	1	7	4	7	13	2	10	4	9	2
19	POTOMAC RIVER BASIN	5	4	1	5	4	2	7	12	7	10	2	3	3	8	1	4	11	4	2	3	2	6	3	10	3	8	7
20	RAPPAHANNOCK AND YORK RIVER BASINS	16	16	7	15	8	3	15	10	21	15	17	11	16	19	13	17	8	7	6	16	10	6	7	10	3	13	7
21	JAMES RIVER BASIN	10	2	5	13	2	4	5	3	12	10	8	4	10	4	9	9	7	7	4	8	4	17	4	10	11	12	7

TABLE 40
RANKING OF NAR AREAS BY SIZES OF THEIR NEEDS PER CAPITA

The rank shown is the highest that the Area achieved for that need during any planning period, including 1980, 2000 and 2020, and the present.

		Publicly Supplied Water	Industrial Self-Supplied Water	Rural Water Supply	Agricultural Irrigation Water	Power Plant Cooling, Fresh Water Withdrawal	Power Plant Cooling, Fresh Water Consumption	Hydroelectric Power Generation	Commercial Navigation	Recreational Boating	Water Recreation Visitor Days	Sport Fishing Man-Days	Hunting Man-Days	Nature Study Man-Days	Water Quality Maintenance	Flood Damage Reduction, Upstream	Flood Damage Reduction, Mainstream	Flood Damage Reduction, Tidal and Hurricane	Drainage, Cropland	Erosion, Agriculture	Erosion, Urban	Erosion, Streambank	Visual and Cultural, Maintenance of Unique Natural Areas	Visual and Cultural, Maintenance of Unique Shorelines	Visual and Cultural, Maintenance of High Quality Areas	Visual and Cultural, Maintenance of Diverse Areas	Visual and Cultural, Development of Metropolitan Amenities	Visual and Cultural, Development of Quality Areas
1	ST. JOHN RIVER BASIN	20	4	1	1	1	1	3	19	3	12	4	3	8	2	17	11	11	3	2	7	9	1	12	10	12	13	7
2	PENOBSCOT RIVER BASIN	6	1	6	9	2	3	3	5	1	7	4	1	3	1	16	13	11	5	10	2	6	2	12	1	12	13	7
3	KENNEBEC RIVER BASIN	4	4	5	2	1	4	1	17	1	4	4	4	5	4	7	5	11	2	7	3	4	18	12	2	11	13	7
4	ANDROSCOGGIN RIVER BASIN	17	2	13	3	4	5	3	19	5	10	3	6	2	3	7	2	11	8	12	4	8	8	12	3	12	13	7
5	MAINE COASTAL BASINS	18	2	3	10	15	13	8	8	2	2	1	5	11	5	15	18	9	4	6	1	10	4	1	4	12	13	7
6	SOUTHERN MAINE AND COASTAL NEW HAMPSHIRE	14	13	12	9	16	16	9	1	2	2	2	8	14	12	2	15	10	13	14	8	18	7	4	5	12	13	7
7	MERRIMACK RIVER BASIN	4	14	17	8	3	5	10	19	7	1	9	12	4	14	1	3	11	17	16	5	17	10	12	9	12	1	3
8	CONNECTICUT RIVER BASIN	6	7	14	5	4	3	4	13	9	7	11	13	1	9	3	4	11	11	11	5	15	6	12	7	12	13	7
9	SOUTHEASTERN NEW ENGLAND	2	18	20	11	19	19	17	10	8	16	15	19	16	16	13	17	6	20	20	19	20	14	10	10	12	9	5
10	THAMES AND HOUSATONIC RIVER BASINS	3	17	12	17	13	12	7	9	3	4	17	18	6	19	4	12	1	17	17	15	19	18	6	10	8	3	1
11	LAKE CHAMPLAIN AND ST. LAWRENCE RIVER DRAINAGE	13	6	3	3	10	2	1	13	9	6	5	2	7	10	11	1	11	1	4	10	1	3	12	6	5	13	7
12	HUDSON RIVER BASIN	13	11	6	4	3	4	3	7	11	10	20	17	20	7	8	6	11	10	6	13	14	8	12	5	2	4	7
13	SOUTHEASTERN NEW YORK METROPOLITAN AREA	1	21	20	15	19	19	19	12	11	19	18	21	2	9	21	20	3	21	21	21	21	18	9	10	8	11	7
14	NORTHERN NEW JERSEY	1	17	18	19	15	15	13	3	15	17	21	19	20	6	19	4	11	19	19	19	12	18	12	10	9	8	6
15	DELAWARE RIVER BASIN	4	6	14	2	5	4	10	5	16	15	19	11	4	6	17	12	8	14	12	14	9	16	11	10	3	2	7
16	COASTAL NEW JERSEY	8	20	18	1	19	19	19	16	16	14	2	15	1	8	20	21	1	14	16	12	11	11	7	10	12	6	1
17	SUSQUEHANNA RIVER BASIN	7	10	7	6	1	2	4	18	18	10	13	8	9	17	9	6	11	8	3	6	2	11	12	10	6	3	7
18	CHESAPEAKE BAY AND DELMARVA PENINSULA DRAINAGE	8	12	6	3	14	13	18	3	11	21	11	15	17	18	1	18	4	3	11	10	13	15	3	10	4	6	2
19	POTOMAC RIVER BASIN	10	14	10	10	7	3	13	15	19	20	14	11	4	20	7	9	11	10	7	15	4	13	8	10	5	9	7
20	RAPPAHANNOCK AND YORK RIVER BASINS	20	8	2	6	7	1	13	6	17	8	9	3	2	10	2	15	5	2	1	5	4	5	2	10	1	13	7
21	JAMES RIVER BASIN	11	5	7	16	1	1	7	2	17	15	12	10	7	7	10	9	7	11	5	13	3	16	5	10	12	11	7

TABLE 41
RANKING OF NAR AREAS
BY GROWTH RATES OF THEIR NEEDS

The rank shown is the highest that the Area achieved for that need's growth rate during the 1980 to 2020 planning periods.

		Publicly Supplied Water	Industrial Self-Supplied Water	Rural Water Supply	Agricultural Irrigation Water	Power Plant Cooling, Fresh Water Withdrawal	Power Plant Cooling, Fresh Water Consumption	Hydroelectric Power Generation	Commercial Navigation	Recreational Boating	Water Recreation Visitor Days	Sport Fishing Man-Days	Hunting Man-Days	Nature Study Man-Days	Water Quality Maintenance	Flood Damage Reduction, Upstream	Flood Damage Reduction, Mainstream	Flood Damage Reduction, Tidal and Hurricane	Drainage, Cropland	Erosion, Agriculture	Erosion, Urban	Erosion, Streambank	Visual and Cultural, Maintenance of Unique Natural Areas	Visual and Cultural, Maintenance of Unique Shorelines	Visual and Cultural, Maintenance of High Quality Areas	Visual and Cultural, Maintenance of Diverse Areas	Visual and Cultural, Development of Metropolitan Amenities	Visual and Cultural, Development of Quality Areas
1	ST. JOHN RIVER BASIN	11	1	6	1	9	1	2	19	17	18	18	11	17	5	10	14	11	1	10	13	1	6	12	9	11	2	7
2	PENOBSCOT RIVER BASIN	6	4	18	6	10	12	11	2	14	17	21	21	21	4	12	17	11	5	17	18	5	9	12	6	11	2	7
3	KENNEBEC RIVER BASIN	19	16	10	2	1	1	12	9	2	19	2	10	11	13	18	15	11	2	12	19	8	12	12	1	10	2	7
4	ANDROSCOGGIN RIVER BASIN	20	19	15	4	1	1	14	19	21	21	5	17	16	12	9	13	11	4	3	20	8	12	12	7	11	2	7
5	MAINE COASTAL BASINS	18	6	7	8	1	1	18	1	20	20	20	19	20	6	13	20	10	3	19	21	1	8	1	7	11	2	7
6	SOUTHERN MAINE AND COASTAL NEW HAMPSHIRE	10	13	9	16	1	1	18	6	7	15	11	13	12	18	6	5	5	11	9	10	21	12	11	1	11	2	7
7	MERRIMACK RIVER BASIN	16	12	20	18	21	18	8	19	8	12	9	8	8	19	5	10	11	8	2	4	6	12	12	1	11	2	1
8	CONNECTICUT RIVER BASIN	13	10	13	11	15	15	13	14	5	14	3	2	2	10	3	8	11	9	13	5	8	7	12	1	11	2	7
9	SOUTHEASTERN NEW ENGLAND	9	17	19	14	18	19	18	5	3	16	12	12	9	20	8	11	7	12	11	7	20	5	11	9	11	2	1
10	THAMES AND HOUSATONIC RIVER BASINS	8	9	17	20	1	1	5	4	1	10	7	7	6	21	2	7	4	10	8	2	8	12	11	9	1	1	1
11	LAKE CHAMPLAIN AND ST. LAWRENCE RIVER DRAINAGE	15	18	12	3	12	11	15	16	15	13	19	20	19	17	19	12	11	7	4	14	8	10	12	1	1	2	7
12	HUDSON RIVER BASIN	5	8	14	7	11	10	6	11	4	3	17	18	18	9	14	16	11	6	16	16	8	12	12	9	1	2	7
13	SOUTHEASTERN NEW YORK METROPOLITAN AREA	21	14	21	21	18	19	16	13	6	9	15	14	14	11	21	19	9	21	21	17	1	12	11	9	11	2	7
14	NORTHERN NEW JERSEY	3	21	16	19	1	1	7	15	9	6	16	16	15	7	11	18	11	13	15	8	8	12	12	9	1	2	1
15	DELAWARE RIVER BASIN	17	3	5	15	17	13	10	10	11	8	10	4	5	3	16	9	6	14	5	6	8	12	11	9	1	2	7
16	COASTAL NEW JERSEY	1	2	8	13	18	19	16	3	12	2	13	9	10	14	17	21	8	15	18	12	8	11	11	9	11	2	1
17	SUSQUEHANNA RIVER BASIN	7	7	11	5	14	14	9	12	10	7	8	6	7	16	1	4	11	18	7	11	7	12	12	9	1	2	7
18	CHESAPEAKE BAY AND DELMARVA PENINSULA DRAINAGE	14	5	3	9	1	1	18	8	13	5	14	15	13	15	20	6	3	16	1	1	8	11	11	9	1	2	1
19	POTOMAC RIVER BASIN	2	11	4	10	13	17	3	17	18	1	1	1	1	8	4	1	11	20	6	3	8	11	11	9	1	2	7
20	RAPPAHANNOCK AND YORK RIVER BASINS	4	20	1	17	1	1	1	7	19	4	4	3	3	1	15	2	1	19	20	15	4	11	11	9	2	2	7
21	JAMES RIVER BASIN	12	15	2	12	16	16	4	18	16	11	6	5	4	2	7	3	2	17	14	9	8	11	11	9	11	2	7

TABLE 42
KEY DEVICES IN EACH NAR AREA

		Storage Facilities	Withdrawal Facilities	Wells	Conveyance Facilities (Diversions)	Quality Control, Temperature (Power Plant Cooling)	Quality Control, Chemical/Biological	Monitoring Facilities	Flood Plain Management	Local Flood Protection	Watershed Management	Erosion Protection	Drainage Practices	Waterway Management	Recreational Boating Facilities	Land Controls	Land Facilities	Habitat Management	Fishways	Stocking	Insect Control	Research	Education	Water Demand and Allocation Changes	Project Operation Changes
1	ST. JOHN RIVER BASIN																								
2	PENOBSCOT RIVER BASIN																								
3	KENNEBEC RIVER BASIN						X																		
4	ANDROSCOGGIN RIVER BASIN						X									X									
5	MAINE COASTAL BASINS																								
6	SOUTHERN MAINE AND COASTAL NEW HAMPSHIRE																								
7	MERRIMACK RIVER BASIN																								
8	CONNECTICUT RIVER BASIN																								
9	SOUTHEASTERN NEW ENGLAND						X					X													
10	THAMES AND HOUSATONIC RIVER BASINS																								
11	LAKE CHAMPLAIN AND ST. LAWRENCE RIVER DRAINAGE																	X							
12	HUDSON RIVER BASIN																								
13	SOUTHEASTERN NEW YORK METROPOLITAN AREA						X																		
14	NORTHERN NEW JERSEY																								
15	DELAWARE RIVER BASIN						X																		
16	COASTAL NEW JERSEY																								
17	SUSQUEHANNA RIVER BASIN																								
18	CHESAPEAKE BAY AND DELMARVA PENINSULA DRAINAGE																								
19	POTOMAC RIVER BASIN																								
20	RAPPAHANNOCK AND YORK RIVER BASINS																								
21	JAMES RIVER BASIN																								

TABLE 44
NEEDS IN EACH NAR AREA LIKELY TO HAVE LARGE BENEFITS WHEN FULFILLED

		Publicly Supplied Water	Industrial Self-Supplied Water	Rural Water Supply	Irrigation Water	Power Plant Cooling	Hydroelectric Power Generation	Commercial Navigation	Recreational Boating	Water Recreation	Fish and Wildlife	Water Quality Maintenance	Flood Damage Reduction	Drainage Control	Erosion Control	Health	Visual and Cultural
1	ST. JOHN RIVER BASIN											X					
2	PENOBSCOT RIVER BASIN	X	X			X				X	X	X					X
3	KENNEBEC RIVER BASIN	X	X			X						X					
4	ANDROSCOGGIN RIVER BASIN									X		X			X		
5	MAINE COASTAL BASINS							X		X	X						X
6	SOUTHERN MAINE AND COASTAL NEW HAMPSHIRE	X	X			X				X		X					
7	MERRIMACK RIVER BASIN											X					
8	CONNECTICUT RIVER BASIN	X	X							X	X	X					X
9	SOUTHEASTERN NEW ENGLAND	X	X			X						X					
10	THAMES AND HOUSATONIC RIVER BASINS	X						X				X					
11	LAKE CHAMPLAIN AND ST. LAWRENCE RIVER DRAINAGE																
12	HUDSON RIVER BASIN											X					X
13	SOUTHEASTERN NEW YORK METROPOLITAN AREA	X										X					
14	NORTHERN NEW JERSEY											X	X				
15	DELAWARE RIVER BASIN		X			X		X		X	X	X					X
16	COASTAL NEW JERSEY	X							X	X	X	X					X
17	SUSQUEHANNA RIVER BASIN					X				X	X	X					X
18	CHESAPEAKE BAY AND DELMARVA PENINSULA DRAINAGE							X		X	X	X					X
19	POTOMAC RIVER BASIN	X	X							X							X
20	RAPPAHANNOCK AND YORK RIVER BASINS	X	X		X									X	X		
21	JAMES RIVER BASIN					X		X				X					

TABLE 45
RANKING OF DEVICE COSTS WITHIN
EACH NAR AREA BY SIZES

The rank shown is the highest that the device cost achieved for that Area during the three planning periods, including 1980, 2000 and 2020.

		Storage, Upstream	Storage, Mainstream	Groundwater	Desalting	Interbasin Transfers	Public Water Supply Conveyance and Treatment	Industrial Self-Supplied Water Conveyance	Agriculture Irrigation Water Conveyance	Non-Agriculture Irrigation Water Conveyance	Power Plant Cooling Water	Commercial Navigation	Recreational Boating	Water Recreation	Fishing Access	Secondary Waste Treatment	Advanced Waste Treatment	Combined Sewer Overflows Control and Acid Mine Drainage Control	Upstream Flood Damage Reduction	Mainstream Flood Damage Reduction	Drainage Control	Erosion Control	Visual and Cultural
1	ST. JOHN RIVER BASIN	9	3	3	17	17	7	13	6	12	5	17	14	2	13	1	2	4	10	17	11	8	3
2	PENOBSCOT RIVER BASIN	7	17	6	17	17	5	9	12	12	4	3	10	2	10	1	2	3	7	17	10	6	4
3	KENNEBEC RIVER BASIN	9	15	5	15	15	6	13	7	9	3	15	9	1	10	1	2	3	15	15	9	7	4
4	ANDROSCOGGIN RIVER BASIN	10	16	5	16	16	4	11	9	7	3	16	9	2	10	1	2	3	6	4	11	7	16
5	MAINE COASTAL BASINS	8	18	5	18	18	4	10	15	6	3	4	12	10	8	1	2	3	9	18	10	7	2
6	SOUTHERN MAINE AND COASTAL NEW HAMPSHIRE	8	18	8	18	18	7	13	16	10	2	2	9	3	12	1	3	2	6	18	14	6	4
7	MERRIMACK RIVER BASIN	3	6	11	19	3	6	15	16	10	9	19	14	2	12	1	4	2	8	9	16	6	4
8	CONNECTICUT RIVER BASIN	7	6	7	19	19	8	12	12	9	5	10	13	1	11	1	3	2	10	19	13	5	4
9	SOUTHEASTERN NEW ENGLAND	6	12	11	20	4	6	14	16	10	4	6	11	2	12	1	3	1	14	7	15	1	3
10	THAMES AND HOUSATONIC RIVER BASINS	6	4	10	16	19	3	14	18	11	8	3	11	1	14	2	7	3	8	7	17	4	2
11	LAKE CHAMPLAIN AND ST. LAWRENCE RIVER DRAINAGE	8	5	5	17	17	4	15	10	9	2	17	13	2	12	1	17	3	13	8	8	4	2
12	HUDSON RIVER BASIN	7	7	7	18	18	4	14	11	10	4	5	12	1	16	1	3	3	11	9	14	5	3
13	SOUTHEASTERN NEW YORK METROPOLITAN AREA	15	15	9	15	4	5	11	15	8	15	7	7	3	6	1	2	1	15	6	15	8	3
14	NORTHERN NEW JERSEY	9	8	12	20	4	5	11	19	9	10	7	8	3	13	1	2	2	12	3	15	7	5
15	DELAWARE RIVER BASIN	7	8	10	21	12	6	11	13	9	5	3	13	3	14	1	2	16	10	8	15	5	2
16	COASTAL NEW JERSEY	12	7	13	6	16	13	12	8	9	11	6	5	8	7	1	2	5	16	3	10	2	2
17	SUSQUEHANNA RIVER BASIN	9	5	7	19	19	6	15	12	11	4	19	16	1	13	1	3	19	10	6	13	3	3
18	CHESAPEAKE BAY AND DELMARVA PENINSULA DRAINAGE	6	18	9	2	2	9	8	8	12	11	3	7	5	13	1	5	18	10	6	9	3	1
19	POTOMAC RIVER BASIN	11	5	8	19	19	2	14	15	12	8	19	10	3	13	1	4	19	8	9	16	3	1
20	RAPPAHANNOCK AND YORK RIVER BASINS	8	4	6	18	18	5	12	17	10	18	3	8	7	13	1	2	18	3	4	9	4	1
21	JAMES RIVER BASIN	17	8	10	2	16	4	14	18	9	9	2	13	6	12	1	2	19	6	7	15	3	7

TABLE 46
RANKING OF NAR AREAS
BY SIZES OF THEIR DEVICE COSTS

The rank shown is the highest that the Area achieved for that device cost during the three planning periods, including 1980, 2000 and 2020.

		Storage, Upstream	Storage, Mainstream	Groundwater	Desalting	Interbasin Transfers	Public Water Supply Conveyance and Treatment	Industrial Self-Supplied Water Conveyance	Agriculture Irrigation Water Conveyance	Non-Agriculture Irrigation Water Conveyance	Power Plant Cooling Water	Commercial Navigation	Recreational Boating	Water Recreation	Fishing Access	Secondary Waste Treatment	Advanced Waste Treatment	Combined Sewer Overflows Control and Acid Mine Drainage Control	Upstream Flood Damage Reduction	Mainstream Flood Damage Reduction	Drainage Control	Erosion Control	Visual and Cultural
1	ST. JOHN RIVER BASIN	9	7	5	5	7	16	19	1	20	13	13	20	15	21	19	10	15	7	13	11	17	18
2	PENOBSCOT RIVER BASIN	13	13	19	5	7	19	12	4	20	12	4	15	12	18	14	6	11	4	13	13	19	15
3	KENNEBEC RIVER BASIN	12	13	9	5	7	14	20	3	18	13	13	17	11	19	21	13	10	19	13	9	18	14
4	ANDROSCOGGIN RIVER BASIN	15	13	17	5	7	16	16	5	18	16	13	20	14	16	20	11	11	13	8	16	19	21
5	MAINE COASTAL BASINS	8	13	14	5	7	14	14	6	17	6	6	17	21	16	17	15	14	10	13	12	18	16
6	SOUTHERN MAINE AND COASTAL NEW HAMPSHIRE	6	13	10	5	7	15	16	16	15	5	4	7	12	10	14	17	8	7	13	13	13	12
7	MERRIMACK RIVER BASIN	3	5	12	5	1	9	15	10	12	16	13	14	9	10	10	14	5	4	11	15	10	11
8	CONNECTICUT RIVER BASIN	2	4	4	5	7	8	6	7	8	4	7	14	5	9	8	8	4	5	13	6	7	5
9	SOUTHEASTERN NEW ENGLAND	1	6	6	5	2	7	9	10	3	4	3	5	1	3	8	5	2	7	4	17	1	5
10	THAMES AND HOUSATONIC RIVER BASINS	4	4	7	4	7	3	6	12	7	7	2	7	3	9	13	14	7	3	5	10	2	6
11	LAKE CHAMPLAIN AND ST. LAWRENCE RIVER DRAINAGE	6	9	5	5	7	10	16	3	12	5	13	15	12	13	15	21	9	6	11	1	13	11
12	HUDSON RIVER BASIN	2	1	2	5	7	4	5	2	5	3	5	8	2	13	5	4	6	8	10	2	11	2
13	SOUTHEASTERN NEW YORK METROPOLITAN AREA	20	13	2	5	1	6	9	20	2	20	7	1	3	1	2	1	1	19	1	21	13	4
14	NORTHERN NEW JERSEY	2	2	10	5	1	3	3	15	7	12	4	3	6	15	1	2	3	6	1	11	9	7
15	DELAWARE RIVER BASIN	1	2	4	5	2	2	1	2	1	1	1	9	5	5	1	1	16	3	6	3	3	1
16	COASTAL NEW JERSEY	14	11	16	3	7	21	17	7	14	13	8	2	19	4	11	12	13	19	2	8	7	9
17	SUSQUEHANNA RIVER BASIN	3	1	1	5	7	5	3	2	2	1	13	11	1	6	5	3	17	1	4	2	1	7
18	CHESAPEAKE BAY AND DELMARVA PENINSULA DRAINAGE	1	13	3	2	2	12	2	1	10	15	3	2	13	3	9	12	17	4	6	1	3	1
19	POTOMAC RIVER BASIN	5	1	2	5	7	1	6	6	4	7	13	1	9	2	5	4	17	1	1	4	4	2
20	RAPPAHANNOCK AND YORK RIVER BASINS	10	3	8	5	7	15	13	18	16	20	8	13	19	15	13	6	17	3	5	4	15	12
21	JAMES RIVER BASIN	13	6	6	1	4	3	3	14	10	10	1	9	11	8	4	4	17	1	3	5	6	17

TABLE 47
RANKING OF NAR AREAS BY SIZES
OF THEIR DEVICE COSTS PER CAPITA

The rank shown is the highest that the Area achieved for that device cost during the three planning periods, including 1980, 2000 and 2020.

		Storage, Upstream	Storage, Mainstream	Groundwater	Desalting	Interbasin Transfers	Public Water Supply Conveyance and Treatment	Industrial Self-Supplied Water Conveyance	Agriculture Irrigation Water Conveyance	Non-Agriculture Irrigation Water Conveyance	Power Plant Cooling Water	Commercial Navigation	Recreational Boating	Water Recreation	Fishing Access	Secondary Waste Treatment	Advanced Waste Treatment	Combined Sewer Overflows Control and Acid Mine Drainage Control	Upstream Flood Damage Reduction	Mainstream Flood Damage Reduction	Drainage Control	Erosion Control	Visual and Cultural
1	ST. JOHN RIVER BASIN	1	1	1	5	7	3	6	1	7	4	13	12	4	8	1	2	11	3	13	2	8	8
2	PENOBSCOT RIVER BASIN	8	13	10	5	7	8	1	3	13	2	1	4	2	2	1	1	6	2	13	4	17	2
3	KENNEBEC RIVER BASIN	6	13	2	5	7	1	6	2	3	5	13	3	1	6	7	4	5	19	13	3	14	1
4	ANDROSCOGGIN RIVER BASIN	9	13	3	5	7	5	4	4	2	7	13	9	7	1	6	3	7	3	1	7	17	21
5	MAINE COASTAL BASINS	6	13	3	5	7	3	2	6	1	1	3	7	20	3	3	5	12	6	13	5	16	2
6	SOUTHERN MAINE AND COASTAL NEW HAMPSHIRE	5	13	7	5	7	15	13	15	4	3	1	1	11	1	9	12	9	3	13	12	4	8
7	MERRIMACK RIVER BASIN	1	2	12	5	1	6	15	12	10	16	13	14	6	8	8	13	4	2	5	16	1	5
8	CONNECTICUT RIVER BASIN	3	4	8	5	7	13	9	7	4	6	7	18	2	10	13	10	2	7	13	9	6	6
9	SOUTHEASTERN NEW ENGLAND	2	9	15	5	3	14	19	12	16	11	5	13	7	15	20	15	3	14	7	20	1	14
10	THAMES AND HOUSATONIC RIVER BASINS	4	4	11	4	7	1	10	15	6	9	4	7	1	18	20	18	13	7	3	17	3	9
11	LAKE CHAMPLAIN AND ST. LAWRENCE RIVER DRAINAGE	1	2	3	5	7	5	11	3	1	2	13	11	12	8	11	21	10	6	7	1	5	4
12	HUDSON RIVER BASIN	3	3	7	5	7	7	14	6	7	4	6	9	2	20	12	9	14	13	8	10	15	6
13	SOUTHEASTERN NEW YORK METROPOLITAN AREA	20	13	15	5	2	17	20	20	21	20	10	15	13	5	11	2	1	19	5	21	21	18
14	NORTHERN NEW JERSEY	5	8	18	5	1	11	11	19	20	14	7	7	9	21	4	1	8	10	2	19	17	16
15	DELAWARE RIVER BASIN	2	10	14	5	3	9	4	10	11	7	3	19	12	17	5	6	16	7	10	15	12	9
16	COASTAL NEW JERSEY	10	11	16	3	7	21	19	6	15	13	8	1	20	1	14	14	15	19	1	10	2	7
17	SUSQUEHANNA RIVER BASIN	3	3	5	5	7	10	11	5	9	3	13	19	1	15	15	4	17	5	6	6	3	12
18	CHESAPEAKE BAY AND DELMARVA PENINSULA DRAINAGE	4	13	11	2	2	19	2	5	15	17	4	3	15	9	17	18	17	8	4	3	2	1
19	POTOMAC RIVER BASIN	9	3	10	5	7	1	17	10	16	12	13	5	13	7	18	5	17	4	1	11	10	3
20	RAPPAHANNOCK AND YORK RIVER BASINS	5	1	5	5	7	14	5	16	7	20	3	3	18	11	9	3	17	1	3	2	8	2
21	JAMES RIVER BASIN	17	5	8	1	4	2	8	14	13	10	2	8	15	12	4	7	17	1	4	8	4	20

AREA 9 SOUTHEASTERN NEW ENGLAND

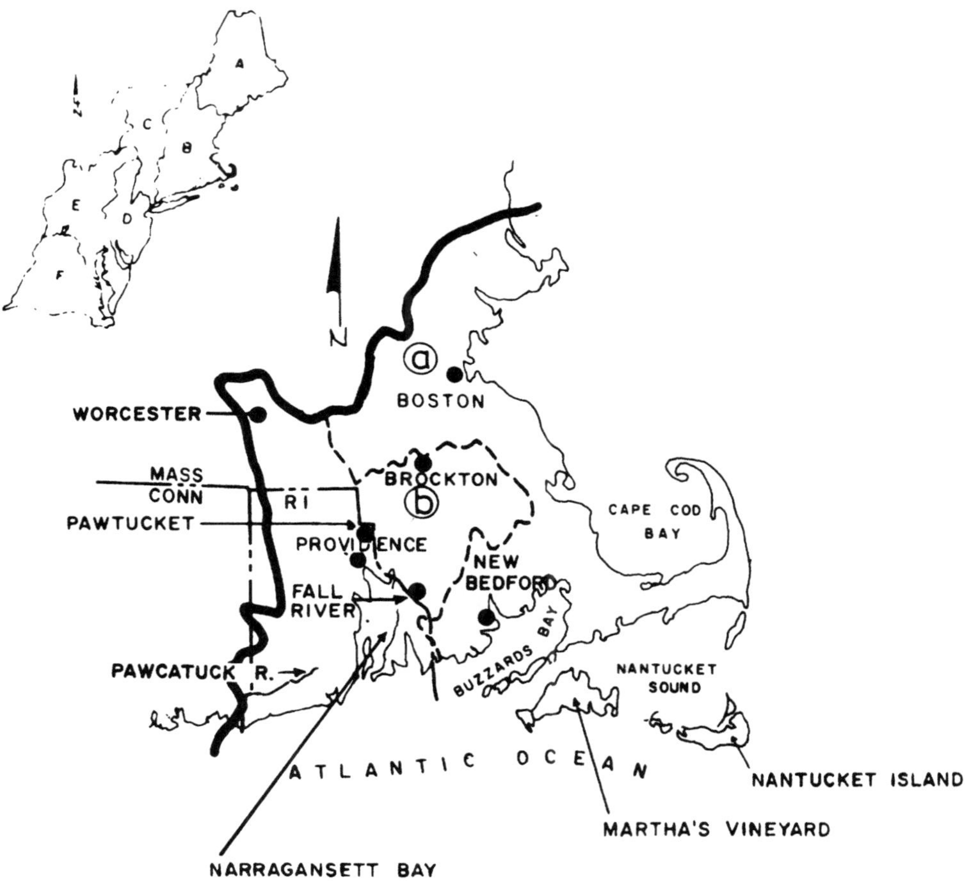

AREA 9

Southeastern New England. Area 9 consists of all Atlantic coastal drainages of Massachusetts (except for the Merrimack River Basin) and Rhode Island. The Area has 4,576 square miles and is divided into two sub-areas. Sub-area a is all of the Massachusetts coastal drainages in the Area. Sub-area b is all of the Rhode Island coastal drainages which also drain a large portion of Massachusetts and a very small portion of Connecticut.

The Area has a unique shoreline and varies from coastal plain to rolling hills. Fine harbors and ports exist along the entire coastline that also has natural scientific interest. These ports include the large commercial ports of Boston and Providence and the very fine recreational harbors that abound along the Area's entire coast especially in Massachusetts, Cape Cod and Narragansett Bays. There are many resort and recreation sites including Cape Cod, Nantucket Island and Newport. A little over half of the Area is rural forest-town with medial visual quality. The rest of the Area is urban - including the Boston, Mass. and Providence, R. I. metropolitan regions. This Area has New England's highest population density.

The Area's very extensive, high quality, but fragile shorelines and natural scientific areas are significant attributes. Rural landscape covers about 60% of the Area and consists entirely of forest-town units of medial visual quality.

Total population was 4.9 million in 1960 with a projected increase to 8 million by 2020. The 1960 density was 974 people per square mile which is expected to increase to 1,833 people per square mile by 2020. The densest part of the Area is Suffolk County, Mass., with 14,207 people per square mile. Sub-area a has 1,222 people per square mile and b has 818 people per square mile.

Area 9 is the commercial, industrial, financial, educational and research center of New England. The 1960 employment of 1.7 million is expected to rise to 3.3 million by 2020. Industries with the largest 1960 employment include services; wholesale and retail trade; machinery; and transportation, communication and public utilities. Employment is expected to decrease in agriculture, forestry and fisheries; food and kindred products; textile mill products; apparel and other textiles; lumber, wood products and furniture; petroleum and coal products; and wholesale and retail trade.

Per capita income was 9% above the 1960 national average, but is expected to decrease to 4% above by 2020.

Water is presently transferred into the Area from the Connecticut and Merrimack River Basins. These inter-basin transfers are expected to increase since there are no major water sources within the Area capable of meeting increasing future needs.

Water pollution in the Area is extensive and sharply limits the usefulness of all of the larger rivers and some portions of the coast. The largest water-using industries in Area 9 are chemicals and plastics, agricultural irrigation and paper. Chemicals and plastics are expected to still be the largest users in 2020, but paper will change to the second largest user followed by primary manufacturing.

Studies are presently being conducted in Area 9 including the Charles River Basin, Massachusetts; the Pawcatuck and Narragansett Bays Study; the Northeastern United States Water Supply Study (NEWS) which has assigned this Area a high priority; and the S. E. New England Comprehensive Study.

Area 9's average annual runoff is approximately 5,280 m.g.d. The existing minimum monthly flow (shortage index 0.01) is 1,150 m.g.d., and the corresponding seven-day minimum is about 60% of this total, or 690 m.g.d. (See Appendix C). The addition of 110 m.g.d. as an allowance for the portion of the consumption losses reflected in the streamflow measurements, results in an existing firm resource available for use of about 800 m.g.d., or 15% of the average runoff. This does not include about 290 m.g.d., which can be imported into the Area by the Metropolitan District Commission.

The practical limit of development within the Area, based on potential yield of new surface storage and additional ground water, would provide a maximum available resource of 1,832 m.g.d., or 35% of the average runoff. Potential sources which would develop the increase of 1,032 m.g.d., include major storage, accounting for 8% of the increase; upstream storage, 20%, and ground water development, 72%.

Possible Alternative Planning Objectives. It is possible to emphasize any of the NAR planning objectives in Area 9 - alone or in combination. There are several characteristics of the Area, however, that should be considered when making a choice. First, Area 9 is one of the most densely populated of the NAR Areas, surpassed only by Area 13. It also includes some of the most extensive coastal resources in the Nation. This combination is a unique situation of extensive recreation and visual and cultural resources being very close to the people who are creating a large demand for their use. Second, sub-area a is highly industrialized, in addition to its large population, and has a great need for good water supplies. Third, sub-area b, while less densely populated and containing less industry, has income and employment problems. Fourth, Regional Development emphasis in sub-area a would be very expensive because of the lack of water in the sub-area for expansion of services. Fifth, Environmental Quality can be emphasized a great deal in Area 9. Such an emphasis would also be expensive but it could make a large change in the urban and rural environmental amenities.

Recommended Mixed Objective. Two sets of mixed objectives are recommended for Area 9 due to the significant differences between the income and employment levels of sub-areas a and b. It is recommended that sub-area a primarily emphasize Environmental Quality with some attention given to National Efficiency. Sub-area a will be able to hold its economic position within the Region without outside aid, but there is a need to preserve its coastal and rural landscapes and improve its urban environment.

Sub-area b will need help in eliminating its unemployment. Its water resources managment program should be oriented towards increasing Regional Development, with some Environmental Quality to begin to retain the natural amenities of the sub-area.

Needs To Be Satisfied Water quality maintenance, publicly supplied and industrial self-supplied water and power plant cooling will be the most important needs in sub area a for gaining the mixed objective but will also be very difficult needs to meet. Publicly supplied and industrial self-supplied water withdrawal needs alone will be the most important to obtaining the mixed objective benefits in sub-area b.

Water pollution is a large problem in sub-area a because of its dense urbanization and high industrial development. Those needs which contribute most towards environmental quality, however, can only be fulfilled in sub-area a if high water quality is maintained. This key action would insure the usefulness of investments in the environmentally oriented needs: recreational boating, water recreation, fish and wildlife and visual and cultural. Water quality maintenance in sub-area b must achieve the state's standards but in a way that allows maximum industrial development combined with some environmental quality.

Large increases in population and industry in both sub-areas will be the cause of the large water withdrawal needs. Publicly supplied water needs will almost triple by 2020 and industrial self-supplied water needs will more than triple by 2020. Commercial navigation needs are large, grow rapidly and are fairly important to the industrial growth of both sub-areas especially during the later years of the planning period. While hydroelectric power generation will decrease during the planning period, reaching zero in 2000, the power plant cooling need will increase tremendously to match the industrial growth of the Area and the availability of coastal sea water. Each of the first two planning periods should realize more than a doubling of power requirements and a 50 percent increase between 2000 and 2020.

Agricultural irrigation needs are not of great importance to the Area but are large and will have a large increase in the first planning period as the industry increases its efficiency to remain competitive. This need will level off during the last two planning periods as growing urban areas prevent it from any expansion and as the change to irrigation is completed. Non-agricultural irrigation needs will grow slowly throughout the period of the study, as golf courses expand to meet the population's recreation needs. The rural water supply need is comparatively small in Area 9 and it decreases during the third planning period as more rural inhabitants go onto publicly supplied water systems.

Water recreation and recreational boating needs are very large and important in Area 9. While the Area has some of the most extensive natural coastal waterways to fill these recreation needs, the facilities to insure proper use and preservation of these resources require a great deal of improvement. Even with the meeting of present water quality standards, which is central to obtaining maximum net benefits to this Area's mixed objective, the investment will be very large for water recreation and visual and cultural needs. The fulfillment of both of these latter needs will be important to the attainment of Environmental Quality benefits in sub-area a. Water recreation needs will grow rather steadily during the planning period. Visual and cultural needs, particularly quality landscape development, are very large and should grow most rapidly during the first stages of the planning period to insure protection of the remaining important landscape areas while they are available and to develop large areas of quality landscape. Urban amenities also need to be developed at this time to provide urban environmental quality.

Coastal shoreline and urban erosion needs are large and when fulfilled will return key benefits to the mixed objective of Area 9 from water recreation needs. Damages will be reduced on the coastal shorelines that are caused by hurricanes and storms and reduced in the urban areas that are caused by poor control of planning and new construction. Fulfillment of the large fish and wildlife and water recreation needs are also key to the visual and cultural needs. Commercial fishing will require special attention to retain its ability to contribute to the Area's income and cultural vitality. Meeting water quality maintenance needs and controlling land use around marshes and shorelines are especially important to fish and wildlife needs.

Flood damage reduction needs are large in this Area and they are expected to double during each of the last two planning periods as land becomes scarce during expansion of the Area's industry and population. Tidal flooding from hurricanes and from other coastal storms and flooding related to major urban drainage problems are particularly significant to the Area's development.

Fulfillment of health needs is important in Area 9 including the safeguarding of local shell fishing, extension of the control of mosquitoes and prevention of the spread of encephalitis. Drainage control needs are small and generally not important in this Area.

Devices. Water is an important component of many activities in this Area. Two types of devices are most important for fulfilling the Area's more significant needs: conveyance facilities and quality control facilities. The most important devices for supplying the Area's water withdrawal needs will be conveyance facilities. These facilities will provide inter-basin diversions with some facilities for surface storage and ground water development. Publicly supplied water will be the largest of the needs to be met by these devices, primarily due to a future shift by much of the Area's industry from self-supplied to publicly supplied systems. Similarly, rural populations will continue to shift to central water systems.

The potential for surface water development will be very low in the Area, especially in sub-area a, unless high water quality standards are maintained to allow increased use of river water. Rhode Island, however, expects to be self-sufficient in water through 2020. Ground water is available in both sub-areas in consolidated rocks and glacial deposits but should be carefully developed to minimize quality problems and adverse effects on existing streamflows.

The second type of device includes almost all means of water quality control along with research for additional means of control. These devices will be necessary to meet water quality maintenance needs in sub-area a, as well as all withdrawal and many instream needs for the whole Area, such as recreation, fish and wildlife and visual and cultural. Water quality maintenance as well as irrigation water needs may be partially fulfilled in the future through the use of treated effluents. This device must be studied further.

Continued economic development in the Area, especially in sub-area a, also depends upon two other types of devices. The first are waterway management devices. Channel improvements will be important to meet commercial navigation needs and the same device along with recreational boating facilities will be needed for continued growth of recreational boating. The second set of devices will be those which satisfy power plant cooling needs. Only Saline water will be used in this Area but cooling towers and ponds and non-condensing power facilities will be needed. There may be a future shifting of power plants between the sub-areas depending upon how it is decided to best preserve the environmental quality associated needs of different parts of the Area -- especially the visual and cultural and fish and wildlife needs.

Investments in land control, land facilities and water access and in biological devices are fairly important in sub-area a for the visual and cultural and fish and wildlife needs which are large and difficult to meet. Fish and wildlife needs can not be completely met within the Area.

Devices of a somewhat lesser importance to achieving the mixed objective of this Area include most of those of a Water/Land nature. Shore erosion protection for selective portions of the ocean and river shorelines of the Area will be key to fulfillment of recreation needs. Watershed management activities, although fairly small, will be important in meeting flood damage reduction, drainage control and visual and cultural needs of the Area. Local flood protection will be primarily for fulfilling hurricane and tidal flood damage reduction needs in the Area. Flood plain management will be able to fulfill the rest of the downstream flood damage reduction needs, especially in sub-area a, while some storage and watershed management devices will be required with flood plain management to fulfill the upstream needs.

The Area's important health needs will require insect control and water quality monitoring to aid, respectively, in the reduction of encephalitis and in the safeguarding of shell fishing.

Devices which change the levels of demand can be of great importance in this area for certain needs. Power plant cooling needs will be reduced and hydroelectric power needs will be met in this manner as non-condensing power facilities are used and reduce the need for cooling water. Recreation and fish and wildlife needs will similarly be met by the opening of existing public water supply reservoirs to public access and use. Such reservoir use changes, however, generally require new legislation and existing health laws require that such reservoir water be treated before it is used in public water systems. Pricing and rationing will also be useful in helping to meet fish and wildlife needs of the Area.

Other legal needs include: increased state regulation of well fields for safeguarding rural water supplies; strengthening of international commercial fishing pacts; better regulation of recreation craft; and interstate agreements for water diversions.

In view of the size of this Area's needs and of the central role that water plays in this Area's development, a Type II comprehensive water resources study should be carried out.

Benefits. Diversion of additional water into sub-area a will result in large benefits as it will allow publicly supplied water needs to be met. This action will also have multiple-use benefits since it will result in cost reduction for meeting industrial self-supplied water, irrigation water, water quality maintenance, water recreation and fish and wildlife needs.

Benefits from meeting water quality maintenance needs should also be high for sub-area a. Large multiple-use benefits from this need will accrue to publicly supplied water, fish and wildlife and visual and cultural needs. Any use of effluent irrigation methods will increase benefits even further due to the reuse of the water, and the inexpensive and useful disposal of the wastes.

Meeting commercial navigation needs will result in fairly large benefits to the Area. This need is an integral part of both sub-areas industrial activities and will help fill all the recreational boating needs and reduce the costs in providing recreational boating facilities. The several navigation devices, along with public walking access to harbors will also produce multiple-use benefits to the visual and cultural needs.

Large benefits will accrue to the Area from the use of saline water for power plant cooling and from any future shifting of power plant locations from one part of the Area to another. These changes will increase the benefits for visual and cultural, fish and wildlife, water quality maintenance, publicly supplied water and industrial self-supplied water needs.

The benefits from investments in water recreation, visual and cultural, fish and wildlife, flood damage reduction and non-agricultural irrigation needs, while smaller than those just mentioned, will be fairly important to the Area and especially important to achievement of the Environmental Quality portion of the recommended mixed objective of sub-area a.

Devices that produce changes in demand for water recreation and fish and wildlife needs would provide especially large benefits with little investment. Pricing and rationing are the most likely devices to be used for this purpose.

Fulfillment of industrial self-supplied water needs will provide large and key benefits for the Regional Development objective of sub-area b that depends upon keeping basic resources readily available for any industrial growth. The enhancement of environment caused by fulfillment of shore erosion needs will be key to achievement of water recreation and visual and cultural needs and will increase the benefits that accrue to them.

There will also be benefits from meeting the needs of rural water supply, agricultural irrigation, drainage control, and health. These will be small benefits although watershed management will serve multiple uses aiding in the achievement of flood damage reduction, drainage control, erosion control, visual and cultural and water recreation needs. Fulfillment of urban erosion control needs will produce large benefits but benefits to coastal erosion control will be controversial. Coastal erosion control will be necessary to preserve much of this Area's shoreline but certain portions of the public will intensely fight its use to keep the shorelines natural.

Costs. Water quality maintenance costs, because of the quantity of wastes and the lack of treatment knowledge, are very high particularly in sub-area a where most of the efforts will be centered. Water recreation, erosion control, and visual and cultural costs will also be very high, especially during the earlier years of the planning period. The degree of urbanization and the size of the needs make these needs costly to meet in this Area. Recreation costs are highest in 1980 because of the inclusion of unsatisfied present demands. Visual and cultural needs are highest in 1980 because of the urgency of preserving and developing unique landscapes and shorelines early in the program. There will be great difficulty in controlling coastal erosion and these costs will be the largest in the Region for meeting this need. Agriculture erosion costs are small but relatively important while urban erosion costs are fairly large.

Publicly supplied water and commercial navigation needs will be costly to meet. Initial capital investment costs for publicly supplied water needs will not be large in this Area. The costs will be proportionally larger than the needs they are fulfill in the first planning period, however, as allowances for project expansion are built into the projects and as the projects are built large enough to take full advantage of each site.

Additional costs will arise in this Area due to the interaction of some devices. High levels of nitrates found in some groundwater supplies and the maintenance of a minimum water table in the cranberry bogs will reduce the ability to use all available groundwater. Cranberry bog waters return to local streams with high levels of pesticides increasing water quality maintenance costs. Urban growth pressures will preclude the filling of some of the peak requirements for non-agricultural (golf course) irrigation needs. Some people in the Area will feel that shoreline erosion control, flood damage reduction and recreation projects will conflict with visual and cultural needs.

Alternative Programs. Emphasizing National Efficiency for all of Area 9 would significantly reduce the variety of devices used although water quality standards would still be met. The need and costs for publicly supplied water would stay high in the whole Area. This action would maintain the industry and population in sub-area a and encourage more industry in sub-area b. Diversions into the Area would still be necessary. Water recreation and visual and cultural needs would be greatly reduced in sub-area a. There would be no reduction in recreational boating needs because of the immense quantities of coastal resources available for such uses so close to several large population centers. Power plant cooling needs for saline water withdrawal would be raised slightly and their costs would be eliminated as most of the power plants would be put into sub-area a where the largest demand for power exists. Erosion and agricultural irrigation needs would be reduced as only the more efficient agricultural operations would continue to function. The large coastal erosion costs would be eliminated. Flood damage reduction needs would be fulfilled through increased use of hurricane barriers.

With a Regional Development emphasis in this Area changes would appear primarily in the program of sub-area a. Fewer water quality maintenance devices would be used including control of nutrients, storm water discharges and marine oil spills and separation of combined sewers. The costs would be reduced for power plant cooling as non-condensing facilities would not be used. Agricultural irrigation needs and costs would be raised for additional income to the Area. Water recreation needs for visitor days would rise slightly but the devices and costs used to meet these needs would be greatly lowered to increase the income to the Area through a lower quality of experience. Upstream flood damage reduction would be achieved through less use of flood plain management, especially in the last planning period; less use of watershed management and greater use of upstream reservoirs. More ocean projects would be used to meet the Area's tidal and hurricane flood damage reduction needs. Drainage control needs and costs would be increased, especially for forest, while erosion control needs and costs for streambanks and shorelines would be greatly reduced.

The emphasis upon Environmental Quality would be difficult to increase in this Area. The recommended levels of needs, devices and costs for this emphasis are already high for sub-area a, because of the present and expected levels of population and industrial growth in the region. Cropland drainage control and agriculture irrigation needs and costs can be increased only slightly because they are already large in the recommended program. Drainage control problems are especially large and intra-urban flood control projects would be used in the new and old urban communities. The levels of needs, devices and costs are also high in sub-area b because of the large emphasis in the present program to increase this sub-area's income and employment levels. Tourism can be increased slightly in this sub-area in a manner that, especially along the coast line, could help the local economy. The industry in the sub-area, however, should be encouraged rather than hurt when the water quality standards are enforced.

AREA 9 AREA 9

NEEDS-cumulative	Pres.	MIXED OBJECTIVE 1980	2000	2020	ENVIRONMENTAL QUALITY 1980	2000	2020	NATIONAL INCOME 1980	2000	2020	REGIONAL DEVELOPMENT 1980	2000	2020
Publicly Supplied Water (mgd)	620	800	1130	1770	760	1010	1410	800	1120	1700	800	1130	1770
Industrial Self-Supplied Water (mgd)	180	290	510	690	280	450	570	280	480	630	290	510	690
Rural Water Supply (mgd)	9.2	9.9	15.0	10.0	←	←	←	9.9	15.0	10.0	→	→	→
Irrigation Water: agriculture (1000 afy)	12	33	34	34	36	38	38	31	21	21	36	38	32
non-agriculture (1000 afy)	10	26	43	65	27	44	65	26	43	65	26	43	65
Power Plant Cooling: withdrawal, saline (cfs)	5000	12000	29000	46000	12000	29000	44000	12000	31000	50000	12000	28000	47000
brackish (cfs)													
fresh (cfs)													
consumption, brackish (cfs)													
fresh (cfs)													
Hydroelectric Power Generation (mw)	3	0	0	0	←	←	←	5	0	0	→	→	→
Navigation: commercial (m.tons annually)	43	56	94	159	49	64	83	53	83	132	56	94	159
recreational boating (1000 boats)	190	250	560	910	←	←	←	250	560	910	→	→	→
Water Recreation: visitor days (m.)	x	92	146	212	92	146	212	80	127	199	94	150	212
stream or river (miles)	x	630	820	1170	630	820	1170	210	270	390	310	400	580
water surface (1000 acres)	x	230	340	470	230	340	470	62	93	130	120	180	250
beach (acres)	x	1600	2000	2400	1600	2000	2400	600	800	1000	700	900	1000
pool (m. sq. ft.)	x	27	35	41	27	35	41	11	15	19	14	18	20
land facilities (1000 acres)	x	120	160	200	120	160	200	20	30	40	40	60	70
Fish & Wildlife: sport fishing man-days (m.)	10	12	15	18	←	←	←	12	15	18	→	→	→
surface area, lake (acres)	x	2000	8500	25000	←	←	←	2000	8500	25000	→	→	→
stream (acres)	x	800	2100	3600	←	←	←	800	2100	3600	→	→	→
access, fresh (acres)	x	110	280	470	←	←	←	110	280	470	→	→	→
salt (acres)	x	860	2310	3990	←	←	←	860	2310	3990	→	→	→
anadromous (acres)	x	6	8	10	←	←	←	6	8	10	→	→	→
piers (1000 feet)	x	24	65	113	←	←	←	24	65	113	→	→	→
hunting man-days (m.)	1.5	1.7	2.0	2.4	←	←	←	1.7	2.0	2.4	→	→	→
access (1000 sq. mi.)	x	0.5	1.0	1.2	←	←	←	0.5	1.0	1.2	→	→	→
nature study man-days (m.)	5.9	6.9	8.6	10.5	←	←	←	6.9	8.6	10.5	→	→	→
access (1000 acres)	x	11	28	49	←	←	←	11	28	49	→	→	→
Water Quality Maint.: non-industrial (m. PEs)	4.3	5.3	6.5	8.1	←	←	←	5.3	6.5	8.1	→	→	→
industrial (m. PEs)	3.4	5.7	9.7	16.7	←	←	←	5.7	9.7	16.7	→	→	→
Flood Damage Reduction:													
avg. ann. damage, upstream (m.$)	6	9	17	34	←	←	←	9	17	34	→	→	→
mainstream (m.$)	3	5	9	18	←	←	←	5	9	18	→	→	→
tidal & hurricane (m.$)	7	11	20	41	←	←	←	11	20	41	→	→	→
Drainage Control: cropland (1000 acres)	6	8	13	14	9	15	16	8	13	14	9	15	16
forest land (1000 acres)	x	0	1	3	0	1	3	0	1	3	1	3	7
wet land (1000 acres)													
Erosion Control: agriculture (1000 acres)	54	70	78	80	70	78	80	64	70	71	70	78	80
urban (1000 acres)	530	680	920	1280	680	920	1280	590	770	1050	680	920	1280
stream bank (mi.)	x	5	14	23	5	14	23	1	3	5	2	7	12
coastal shoreline (mi.)	x	540	1090	1120	540	1090	1120	3	11	19	8	23	38
Health: vector control and pollution control	x	x	x	x	x	x	x	x	x	x	x	x	x
Visual & Cultural:													
landscape maintenance, unique natural (sq.mi.)	50	850	850	850	850	850	850	850	850	850	Same	as	EQ
unique shoreline (mi.)	x	20	20	20	20	20	20	10	10	10	Same	as	EQ
high quality (sq.mi.)													
diversity (sq.mi.)													
agriculture (sq.mi.)													
landscape development, quality (sq.mi.)	x	200	400	600	200	400	600	100	200	300	Same	as	EQ
diversity (sq.mi.)													
metro. amenities (mi.)													
" " (sq.mi.)	x	50	50	50	50	50	50	0	50	50	Same	as	EQ

DEVICES—incremental	MIXED OBJECTIVE				ENVIRONMENTAL QUALITY				NATIONAL INCOME				REGIONAL DEVELOPMENT			
	Purposes	1980	2000	2020	Purposes	1980	2000	2020	Purposes	1980	2000	2020	Purposes	1980	2000	2020
I. Resource Management																
A. Water																
Storage Facilities ϕ																
reservoirs, upstream (1000 af)	FW,Rec,VC	156*	x*	x*	Irrig ƒ	17	1	x	Irrig ƒ	13	x	x	Irrig ƒ	17	1	x
mainstream (1000 af)	FW,Rec,VC,WQ	26*	x*	x*	PS,WQ ƒ	145	33	59	PS,WQ ƒ	179	43	202	PS,WQ ƒ	185	44	219
Withdrawal Facilities																
intakes & pumping, fresh (mgd)	PS,Ind,Irrig	240	530	680	**	200	370	330	**	230	500	590	**	240	530	680
brackish (mgd)	Ind	70	120	160	Ind	70	90	110	Ind	70	100	140	Ind	70	120	170
estuarine (mgd)																
ocean (mgd)	Pow	x	x	x	Pow	x	x	x	Pow	x	x	x	Pow	x	x	x
wells (mgd)		29 *	10 *	9*	Rur**	87	128	19	Rur**	93	125	20	Rur**	98	121	20
Conveyance Facilities																
interbasin diversions, into (mgd)		120*	490*	740*	PS	63	112	130	PS	80	150	300	PS	80	150	360
out of (mgd)																
Quality Control Facilities																
temperature, cooling towers & ponds	WQ	x	x	x	WQ	x	x	x	WQ	x	x	x	WQ	x	x	x
chemical/biological																
potable water treat plants (mgd)	PS	74	120	520	PS	58	89	227	PS	72	117	450	PS	74	120	520
waste treatment plants																
secondary (85%) (m. PE removed)	WQ,VC	9.4	0	0	←				WQ,VC	9.4	0	0	→			
secondary (90%) (m. PE removed)	WQ	0	15	22	←				WQ	0	15	22	→			
advanced (95%) (m. PE removed)	WQ	0	0.81	1.24	←				WQ	0	0.81	1.24	→			
effluent irrigation	Irrig	x	x	x												
nutrient control	WQ,VC	x	x	x	WQ,VC	x	x	x								
stormwater discharge control	WQ,VC	x	x	x	WQ,VC	x	x	x								
acid mine drainage control																
septic tank control	WQ,VC	x	x	x	WQ,VC	x	x	x								
separate combined sewers	WQ,VC	x	x	x	WQ,VC	x	x	x								
Pumped Storage																
Desalting Facilities																
Monitoring Facilities																
B. Water/Land																
Flood Plain Management																
upstream (1000 acres)	FDR,VC	31	19	84	FDR,VC	34	21	124	FDR,VC	28	15	43	FDR,VC	27	16	43
mainstream (1000 acres)	FDR,VC	x	x	x	FDR,VC	x	x	x	FDR,VC	x	x	x	FDR,VC	x	x	x
Local Flood Protection																
ocean (projects)	FDR	0	3	0	FDR	0	1	0	FDR	0	4	0	FDR	0	4	0
river (projects)	FDR	7.5	9.5	0	FDR	1.0	0	0	FDR	13.0	20.0	0	FDR	14.0	19.0	0
flood control channels (mi.)																
Watershed Management (1000 acres)	FDR,Drn,VC	180	310	190	FDR,Drn,VC	190	390	390	FDR,Drn,VC	160	230	x	FDR,Drn,VC	170	220	x
Erosion Protection, land treatment	Ern	x	x	x	Ern	x	x	x	Ern	x	x	x	Ern	x	x	x
coastal shoreline	Ern,Rec	x	x	x	Ern,Rec	x	x	x	Ern,Rec	x	x	x	Ern,Rec	x	x	x
river shoreline	Ern	x	x	x	Ern	x	x	x	Ern	x	x	x	Ern	x	x	x
Drainage Practices	Drn	x	x	x	Drn	x	x	x	Drn	x	x	x	Drn	x	x	x
Waterway Management																
navigation channel improvement	Nav	x	x	x					Nav	x	x	x	Nav	x	x	x
debris removal																
recreation boating facilities	Rec,Nav	x	x	x	Rec,Nav	x	x	x	Rec,Nav	x	x	x	Rec,Nav	x	x	x

* From the supply model for the following purposes: PS, Ind, Rur, Irrig, Pow.
ϕ Flood control storage not included.
** Also includes the following purposes: PS,Ind,Irrig
ƒ Also includes the following purposes: FW,VC,Rec

AREA 9 AREA 9

DEVICES-incremental (cont.)	MIXED OBJECTIVE Purposes	1980	2000	2020	ENVIRONMENTAL QUALITY Purposes	1980	2000	2020	NATIONAL INCOME Purposes	1980	2000	2020	REGIONAL DEVELOPMENT Purposes	1980	2000	2020
C. Land																
Controls																
fee simple purchase (buying) (sq.mi.)	VC, FW	450	200	200	VC, FW	450	200	200	VC, FW	100	0	0	Same	as	EQ	
fee simple purchase (buying) (mi.)	VC, FW	20	0	0	VC, FW	20	0	0	VC, FW	10	0	0	"	"	"	
purchase lease (sq.mi.)	VC,FW	0	0	0	VC,FW	0	0	0	VC, FW	100	150	100	"	"	"	
easements (sq.mi.)																
deed restrictions (sq.mi.)																
tax incentive subsidy (sq.mi.)	VC, FW	600	0	0	VC, FW	600	0	0	VC, FW	700	0	0	"	"	"	
zoning (sq.mi.)																
zoning (mi.)																
zoning and/or tax inc. subs. (sq.mi.)																
zoning and/or tax inc. subs. (mi.)																
Facilities																
recreation development	Rec, Hlth	x	x	x	Rec	x	x	x	Rec	x	x	x	Rec	x	x	x
overland transportation to facility	Rec	x	x	x	Rec	x	x	x	Rec	x	x	x	Rec	x	x	x
parking and trails	FW	x	x	x	←				FW	x	x	x	→			
site sanitation and utilities	VC	x	x	x	VC	x	x	x	VC	x	x	x	Same	as	EQ	
D. Biological																
Habitat Management, fish	FW	x	x	x	←				FW	x	x	x	→			
wildlife	FW	x	x	x	←				FW	x	x	x	→			
Fishways	FW	x	x	x	←				FW	x	x	x	→			
Stocking, fish	FW	x	x	x	←				FW	x	x	x	→			
wildlife	FW	x	x	x	←				FW	x	x	x	→			
Water Quality Standards Enforcement	FW	x	x	x	←				FW	x	x	x	→			
Insect Control	Hlth	x	x	x	←				Hlth	x	x	x	→			
II. Research	WQ	x	x	x					WQ		x					
III. Education																
IV. Policy Changes																
Water Demand and Allocation Changes																
pricing and rationing	FW	x	x	x												
non-condenser power facilities	Pow		x	x	Pow		x	x								
re-circulation (internal)																
Project Operational Changes																
remove restrictions	Rec,FW	x	x													
remove project	FW	x	x	x	FW	x	x	x	FW	x	x	x	FW	x	x	x
add new project needs	Rec, FW	x	x	x	Rec, FW	x	x	x	Rec,FW	x	x	x	Rec,FW	x	x	x
change project design load	Rec	x	x	x	Rec	x	x	x	Rec	x	x	x	Rec	x	x	x
V. Others																
Upstream Flood Control Storage (1000af)	FDR	17	44	0					FDR	30	41	0	FDR	33	88	0
Ground Water Recharge	Ind	x	x	x												
Use of Waste Thermal Heat	Pow	x	x	x												
Oil Spill Control	FW,VC,WQ	x	x	x												
Flood Skimming	Ind	x	x	x												

AREA 9

FIRST COSTS - incremental ($ million 1970)	MIXED OBJECTIVE			ENVIRONMENTAL QUALITY			NATIONAL INCOME			REGIONAL DEVELOPMENT		
	1980	2000	2020	1980	2000	2020	1980	2000	2020	1980	2000	2020
Water Development Costs:												
storage, upstream	46.7*	0*	0*	0.6	0.1	0	0.4	0	0	0.6	0.1	0
mainstream	9.7*	0*	0*	33	19	21	41	25	52	42	25	56
wells	12.8*	4.4*	3.9*	5.8	7.2	3.7	5.8	7.0	3.9	6.1	6.9	3.8
desalting												
Water Withdrawal and Conveyance Costs:												
inter-basin transfers	0.4*	170.7*	57.9*	26.5	45.0	19.3	32.8	59.0	46.0	34.2	60.0	55.0
public water supply	15	28	95	14	24	49	15	28	84	15	28	95
industrial self-supplied water	0.82	1.74	1.84	0.78	1.16	1.26	0.78	1.50	1.58	0.82	1.74	1.84
rural water supply	x	x	x	x	x	x	x	x	x	x	x	x
irrigation, agriculture	2.04	0.16	0	2.55	0.32	0	1.52	0	0	2.55	0.32	0
nonagriculture	14	14	18	14	13	17	14	14	18	14	14	18
Power Plant Cooling Water	0	30	130	0	57	187	0	0	0	0	18	18
Hydroelectric Power Generation												
Navigation: commercial	20	135	89	0	0	0	20	135	89	20	135	89
recreation	4.2	12.5	14.5	←	←	←	4.2	12.5	14.5	→	→	→
Water Recreation	970	630	670	970	630	670	140	100	150	240	120	240
Fish and Wildlife: fishing	4.7	7.5	8.7	←	←	←	4.7	7.5	8.7	→	→	→
hunting	x	x	x	x	x	x	x	x	x	x	x	x
nature study	x	x	x	x	x	x	x	x	x	x	x	x
Water Quality Maint.: waste treatment, secondary	340	780	1180	←	←	←	340	780	1180	→	→	→
advanced	0	170	250	←	←	←	0	170	250	→	→	→
other ≠	1550	0	0	←	←	←	1550	0	0	→	→	→
Flood Damage Reduction: upstream	3.7	6.6	0	0	0	0	8.0	13.3	0	7.4	13.1	0
mainstream	32	27	0	32	7	0	32	30	0	32	30	0
Drainage Control	0.18	0.49	0.18	0.27	0.58	0.18	0.18	0.49	0.18	0.31	0.62	0.27
Erosion Control	973	1013	113	973	1013	113	15	44	59	39	65	85
Health	x	x	x	x	x	x	x	x	x	x	x	x
Visual and Cultural	439	70	70	439	70	70	208	53	35	Same	as	EQ
Summation of Available Estimated Costs	4400	3100	2700	4400	2900	2600	2400	1500	2000	2800	1600	2200

AREA 9

*From the supply model and includes OMR costs.

≠ Combined sewer overflows control and acid mine drainage control.

Chapter 10: Perspectives on the Study

The NAR framework study reported on in this book is important for its contributions to multiobjective planning, including the environmental objective; for the use of mathematical modeling methods; and for the use of institutional and organizational measures to shape the planning process. This chapter provides perspectives on the NAR study as a whole: the results, influence, and lessons of the study. These are dealt with under four headings: the use of the study; forecasts and recommendations; the study as a framework study; and methods and models. (On methods and models also see the perspective sections in Chapters 1-8.)

10.1 THE USE OF THE STUDY

The NAR study has seen extensive use as a guide for some regions and purposes and as a source of methods. A few of many examples are given here. The use most in line with the framework idea was the employment of the demand model to produce forecasts in the U.S. Water Resources Council's Second National Assessment of 1975. (The National Assessment program superceded the framework planning program.) For this assessment, the demand model was rerun to provide forecasts of water needs for the Mid-Atlantic region (Schilling, n.d.). The methods of the NAR also influenced its regional companion plan, the Northeastern United States Water Supply (NEWS) Study. This influence included the use of the multiobjective approach to characterize alternative plans (for example, U.S. Army Corps of Engineers, 1975, ch. 4), and the application of other methods such as the demand model regression equation (Anderson-Nichols, 1971, p. 33). As another example, the NAR study's region 9 (eastern Massachusetts, most of Rhode Island, and a small portion of Connecticut) formed the basis of the New England River Basins Commission's Southeastern New England Study.

The conceptual basis of the visual and cultural ranking system was used in somewhat different form in the Souris-Red-Rainy basins study (Souris-Red-Rainy River Basins Commission, 1972, pp. I-164ff). The study also influenced the multiobjective Principles and Standards of 1973 (U.S. Water Resources Council, 1973) both in general, as an illustration of the feasibility of the approach embodied in the standards, and more specifically, as in the required development of at least one alternative plan for each objective. Illustrative of the influence of the methods of the study are the many scientific papers based on its planning methods, including multiobjectives, the use of mathematical models, and the institutional methods used, of which a sampling have been cited in this book.

Perhaps the most striking aspect of the NAR study as a source of guidance for planning is that a basin or framework study begun today might well employ approaches very similar to those used in the NAR study, adapted to today's computers. The fundamental reason for the longevity of the NAR's methods is that the study was undertaken soon after the development of a new theoretical paradigm for water resources planning, multiobjective theory, and soon after the use of the digital computer became practical for large-scale water planning. The NAR study embodies many of the most important characteristics of the era of water planning which commenced in the early 1960's, the paradigm and techniques of which remain central to the field today.

10.2 FORECASTS AND RECOMMENDATIONS

The NAR plan includes a wide range of forecasts and recommendations, the most important of which have been presented in Chapter 9. These forecasts and recommendations, as with those of any plan, have sometimes been accurate and timely in retrospect, and sometimes not. (A thoughtful comparison of forecast and outcome is provided in Netzer, 1985. He compares the well-known Harvard forecasting study of the New York metropolitan area, the final results of which were published in 1960 for the Regional Plan Association, with 1980's actuality.)

Among the many forecasts and recommendations of the NAR that were accurate and timely were, for example, the forecast of increasing suburbanization of office activities in the region, with its implications for water supply and sewage systems (Report, 47), and the plan's call for a substantial program to maintain high quality landscape areas in the densely populated NAR region (Report, 195).

Many other forecasts and recommendations appear in retrospect to be inaccurate. One such is the forecast of substantially increased use of power cooling water (most of it saline or brackish) in the region, to a level of 84,000 cfs in 1980 from 45,000 cfs at the time of planning. This forecast was too high, to judge from available reports from area utilities to the U.S. Department of Energy on Form EIA-767 (U.S. Department of Energy, 1984). In this case, the planners did not foresee the substantial impact on the U.S. energy sector of the real increases in petroleum prices that occurred following the completion of the plan. As a result, recommendations for supply increases in the NAR plan, insofar as they related to increased power cooling needs, were in the event inappropriate.

As for devices to supply demands for water, the NAR plan recommended incremental storage in mainstem reservoirs of 340,000 acre-feet (Report, 180) to avoid shortages. This was to include the assumed completion of several reservoirs, among them the large Tock's Island reservoir (Report, 182). Yet there have been no substantial mainstem storage facilities constructed or initiated in the NAR since the time of the plan except Gathright Dam in Virginia and North Branch of the Potomac Dam in Maryland and West Virginia, completion of which was assumed. At the same time there have been severe water shortages in the NAR since the plan, notably in the New York and New Jersey areas. This, however, does not mean necessarily that the exact amount of storage recommended in the NAR plan ought to have been provided, but only that some appropriate combination of management devices, additional storage, and the willingness of consumers to bear occasional shortages is required.

Any recommended plan, which is essentially a point forecast, is nearly certain to be both "right" and "wrong" in different respects. These convergences and divergences of forecasts and recommendations from actuality do not in themselves indicate that a plan is adequate or inadequate. The question is whether the planning process as a whole was a useful mechanism for aiding decisions. Here, two things seem clear. First, the idea of alternative plans is a good one. These permit the illustration both of alternative possibilities, and of alternative preferences among objectives. In this regard, perhaps the alternative plans in the NAR study could have been more fully elaborated, given additional resources; and perhaps one or two more alternative plans could have been prepared.

Second, in addition to alternatives in the planning process itself, there is a need for a continuous planning process (Schwarz and Major, 1971, p. 6.; Kamarck, 1983, p. 129). In order to take account of changing possibilities and preferences, there needs to be a planning structure that will permit suitable revisions of the alternatives at appropriate times. (The Report recommends updating the NAR plan every 10 years in the year immediately following each census year; Report, 223.) Such revisions will become increasingly practical with increasing computer storage and better software capabilities.

10.3 THE STUDY AS A FRAMEWORK STUDY

The idea of the framework studies was to provide a regional planning context within which detailed, consistent plans for water resources management and development could be implemented. However, although widely used as a guide for regions and purposes and as a source of methods, the NAR study was not consistently used as a framework plan as originally intended by the U.S. Water Resources Council. (Thus, the NAR study does not provide a complete test of the value of framework planning as a method. For this purpose, a study of the framework plans (Rahmenpläne) and their results undertaken in the Federal Republic of Germany would be helpful.)

A first reason for this, affecting the water resources sector as a whole in the United States, is that Federal funding for water resources development largely came to a halt from shortly after the completion of the NAR plan to 1985 because of controversies over local funding for projects and environmental considerations. This lack of funding drained much of the dynamism from the planning enterprise as a whole, including the framework plans and specifically including work on multiobjectives and mathematical models.

A second reason is that funding for the U.S. Water Resources Council, under whose aegis the framework plans were undertaken, ceased in the early 1980's. While this was the result of factors largely unrelated to the framework plans in themselves, the end of Council activities removed a primary institutional setting for the plans.

A third reason is specific to the framework planning effort. Institutional mechanisms and funding to implement the framework plans after their completion were not fully provided at the time the framework planning idea was developed, and thus there was no automatic or simple way available to insure their use. A lesson that can be drawn from this is that the planners might have employed some of their resources and time during the planning period to begin to develop the institutional mechanisms and the concern among agency personnel that would have been required to use the NAR study as a consistent framework for water planning for the entire NAR region. One method of doing this would be to create and staff, during the study, a small continuing work group to serve as a source of post-study guidance to planning agencies on the methods and conclusions of the study. (See also the suggestion in Major and Lenton, 1979, p. 227, regarding the importance of continued post-study liaison between planners and the sponsoring agency.)

10.4 METHODS AND MODELS

With respect to multiobjectives, the three objectives used in the study, national income, regional development, and environmental quality, appear to represent well the objectives relevant to water and related land planning in the NAR. The extent of iteration about preferences toward objectives among staff and Coordinating Committee representatives seems also to have been generally adequate (although the Coordinating Commmittee's conclusions reported in Section 9.6 of this book are relevant here). The level of detail at which the multiobjective approach was used in the study could have been usefully increased. Even with the constraints on the planners' abilities to handle large amounts of information, given the limited computational capabilities available at the time of planning, it would have been appropriate to attempt further detailed assessment of multiobjective impacts, particularly in the areas of national income benefit assessment and the measurement of regional impacts. Certainly given the revolution in computational capabilities since the NAR plan, more detailed assessments of impacts would be expected in a multiobjective framework plan undertaken today. These same capabilities will also permit extensive exploration of the output formats of alternative plans.

With respect to models, the three principal computer models used in the study, the demand, supply, and storage-yield models, deal respectively with demand (or needs) forecasting, the analysis of supply alternatives, and the assessment of the stochastic physical resource. The specific models chosen appear reasonably appropriate for these purposes, although many variants would have been possible. The models were all developed at a relatively early stage in the application of computer methods to water resources. With the availability of increased computational power, more detailed models could have been chosen and developed. To some extent the development of more detailed models would change the nature and role of the framework plan because more detailed models would approximate standard river basin planning to a greater degree than did the NAR models. Nonetheless, the need for a framework approach would have remained in that the linking of basins, even with more detailed models, would be important in a multi-basin region such as the NAR. An alternative modeling approach to that actually used would therefore be a series of models for the various basins, linked to study the problem of regional consistency.

A substantial amount of time was needed to settle on a suitable set of planning methods and models. The demand and supply models were begun several years after the study was initiated, and they were fully operational only quite far into the study period. Part of this was due to the changing theories and capabilities of the time. The study began in 1966, only four years after the publication of a seminal work on water planning (Maass et al., 1962) from which many of the methods were adopted. Beyond this special factor, however, there is perhaps a general lesson to be learned, which is that with the possible exception of standard studies of a detailed kind there should be time allocated in a study process for the exploration of methods and models. It should be recognized that there will be false starts and changes of direction and emphasis; in any planning situation the models and methods used need to arise out of the necessities of the work. (For an example of changes in modeling methods during the course of a study, see Major and Lenton, 1979, App. A.)

The models did not incorporate all of the multipurposes of planning, for example navigation and flood control, and there were institutional and conceptual problems in linking these to the purposes such as industrial water supply that were included in the models. The planning process as a whole would have been improved had more been done to incorporate one purpose, water quality, into the models. The difficulties encountered in doing this in the NAR study were in part due to the institutional stress at the time of what is now the Environmental Protection Agency; during the planning period the relevant office was part of, successively, the Public Health Service, the Federal Water Pollution Control Administration, and the Environmental Protection Agency. However, another reason was simply that water quality was a relatively new consideration in framework planning at the time, and the range of problems of integrating any new factor into a complex process were encountered with respect to it.

With respect to scope, the study was a sectoral study, focussing on water and related land resources. The study was linked to more general demographic and economic trends through the projections in Appendix B, Economic Base, and was multisectoral in some details. For example, with respect to transportation, driving time from major metropolitan areas was an integral part of the visual and cultural needs assessment (N-100 and Figure 4-4 of this book). The question of multisectoral links deserves more investigation in the design of water resource plans. Given the restriction of the study to water and related land resources, the professional disciplines and the agencies involved appear to have been appropriate (see the list given in Schwarz, Major and Frost, 1975, p. 251).

The scope of the study might have been expanded to include more public involvement, which was at the time of planning beginning a period of rapid growth. One advantage of a substantial public involvement program in the NAR study might have been in eliciting comments on the appropriateness of model results from persons familiar with particular areas. It might also have been true that increased public involvement would have helped to increase the emphasis on the relatively new purpose of water quality planning and to facilitate its integration into the planning process.

REFERENCES

Anderson-Nichols Company, Inc., "Water Deficits for Urban Metropolitan Areas," Contract No. DACW 52-71-C-0001, Hartford, Conn., May, 1971.

Kamarck, Andrew M., Economics and the Real World, Basil Blackwell Publisher Limited, Oxford, England, 1983.

Maass, Arthur, et al., Design of Water-Resource Systems, Harvard University Press, Cambridge, Mass., 1962.

Major, David C., and Roberto L. Lenton, Applied Water Resource Systems Planning, Prentice-Hall, Englewood Cliffs, N.J., 1979.

Netzer, Dick, "1985 Projections of the New York Metropolitan Region Study," American Economic Review, Papers and Proceedings 75:2, May, 1985, 114-119.

Schilling, Kyle E., "1970 NAR Demand Model as Adapted for Use in the 1975 National Assessment," U.S. Army Engineer Institute for Water Resources, Fort Belvoir, Va., n.d.

Schwarz, Harry E., and David C. Major, "An Experience in Planning: The Systems Approach," Water Spectrum III:3, Fall, 1971, 29-34.

Schwarz, Harry E., David C. Major, and John E. Frost Jr., "The North Atlantic Regional Water Resources Study," in J. Ernest Flack, ed., Proceedings of the Conference on Interdisciplinary Analysis of Water Resource Systems, American Society of Civil Engineers, New York, 1975, 245-271.

Souris-Red-Rainy River Basins Commission, Souris-Red-Rainy River Basins Comprehensive Study: Recreation and Preservation, Vol. 5, App. 1, 1972.

U.S. Army Corps of Engineers, North Atlantic Division, Northeastern United States Water Supply (NEWS) Study, Washington Metropolitan Area Water Supply Study: Report, New York, November, 1975.

U.S. Department of Energy, "1984 Form EIA-767, Average Annual Rate of Cooling Water Withdrawal, Discharge, and Consumption," computer printout by State, 1984.

U.S. Water Resources Council, "Water and Related Land Resources: Establishment of Principles and Standards for Planning," Federal Register 38:174, 1973, 24778-24869.

INDEX

Water Science and Technology Library

1. A.S. Eikum and R.W. Seabloom (eds.): *Alternative Wastewater Treatment.* Low-Cost Small Systems, Research and Development. Proceedings of the Conference held at Oslo, Norway (7–10 September 1981). 1982
ISBN 90-277-1430-4
2. W. Brutsaert and G.H. Jirka (eds.): *Gas Transfer at Water Surfaces.* 1984
ISBN 90-277-1697-8
3. D.A. Kraijenhoff and J.R. Moll (eds.): *River Flow Modelling and Forecasting.* 1986 ISBN 90-277-2082-7
4. World Meteorological Organization (Geneva) (ed.): *Microprocessors in Operational Hydrology.* Proceedings of a Conference held in Geneva (4–5 September 1984). 1986
ISBN 90-277-2156-4
5. J. Nemec: *Hydrological Forecasting.* Design and Operation of Hydrological Forecasting Systems. 1986 ISBN 90-277-2259-5
6. V.K. Gupta, I. Rodríguez-Iturbe and E.F. Wood (eds.): *Scale Problems in Hydrology.* Runoff Generation and Basin Response. 1986
ISBN 90-277-2258-7
7. D.C. Major and H.E. Schwarz: *Large-Scale Regional Water Resources Planning.* The North Atlantic Regional Study. 1990 ISBN 0-7923-0711-9

Kluwer Academic Publishers – Dordrecht / Boston / London

1. [illegible] Stochastic [illegible] *Treatment* [illegible]
Low [illegible] *Proceedings* [illegible]
Conference [illegible]

[illegible]

2. W. [illegible] *Urban* [illegible]

[illegible]

3. [illegible] and A.K. [illegible]
1984 [illegible]

4. Wat[illegible] and [illegible] *Quantitative* [illegible] *Proceedings* [illegible]
Op[illegible] September 1985 [illegible]
September [illegible] 1986.

[illegible]

5. J. Ne[illegible] *Potential* [illegible]
Potentials [illegible] 1986 [illegible]

6. V.K. [illegible] *Hydrology* [illegible]

[illegible]

7. D.C. Ma[illegible]
Planning [illegible]

MIX
Papier aus verantwortungsvollen Quellen
Paper from responsible sources
FSC® C105338

If you have any concerns about our products,
you can contact us on
ProductSafety@springernature.com

In case Publisher is established outside the EU,
the EU authorized representative is:
Springer Nature Customer Service Center GmbH
Europaplatz 3, 69115 Heidelberg, Germany

Printed by Libri Plureos GmbH
in Hamburg, Germany